D0881549

Extra Life

Extra Life

Coming of Age in Cyberspace

David S. Bennahum

BASIC
BOOKS

A Member of the Perseus Books Group

Published by Basic Books,
A Member of the Perseus Books Group

FIRST EDITION

Designed by Rachel Hegarty

Library of Congress Cataloging-in-Publication Data
Bennahum, David S.
    Extra life : coming of age in cyberspace / David S. Bennahum. —
1st ed.
        p.   cm.
    ISBN 0-465-01235-3 (hc)
    1. Computers and civilization.   2. Computer games.   3. Cyberspace.
I. Title.
QA76.9.C66B46   1998
303.48'34—dc21
                                                        98-39251
                                                        CIP

98 99 00 01 02 ❖/RRD 10 9 8 7 6 5 4 3

To my teachers

10,000 points earns an extra life.

—Atari Missile Command
instructions, 1980

Contents

Contents

Prologue:
Pirate's Cove

THE FIRST COMPUTER I EVER OWNED sat in a brown box for ten years. In 1986, just before my high school graduation—after a test of wills that failed to end my mother's marriage—I walked out on her and my stepfather and went to live with my dad. When it became clear I wouldn't be coming back, my mom packed up all my possessions and placed them in a warehouse just about a mile from the Queens side of the 59th Street Bridge, creating a time capsule, a snapshot of my life. There they stayed, left to face the extremes of heat and cold until one April morning in 1996 when I crossed the East River to reclaim them.

It was a clear day and the halls of the warehouse were dim. Silent rows of metal containers stood side by side, padlocked, a halfway house for cast-off treasures no longer a part of their owners' lives, yet with too much history to be consigned directly to the curb as worthless trash. It was a place frozen in time, where children come to find their parents' things and parents come to leave their children's things.

My computer should have been trash, thrown into the dustbin of technological history like any other electronic appliance unable to compete with the latest upgrade, but putting it out with the garbage would have been like putting an old pet to sleep. Yet I had done just this, abandoning my old companion to fate along with the rest of my belongings.

As my mom led me down the hall of the warehouse looking for container 10C4, I wondered, Will it still work? Will the floppy disks still be readable? Will the disk drives turn on, will the computer's silicon microprocessor carry electricity from one logic gate to the next? Or will the printed circuits and solder have degraded by now, destroyed by time and the vagaries of chance—of hot summers and cold winters expanding and contracting one vital part or another, pushing one piece beyond its limit and rendering the machine a useless heap of junk?

My mom stood before the metal door, key in hand, and unlocked the container. The door swung outward and weak light from the hallway crept in. I felt like an archeologist poised at the edge of an ancient tomb. Motes of dust glittered in the air. I flipped on our old flashlight and an intimidating wall of boxes came into view. One by one we pulled them out, navigating among cast-off bed frames, tabletops, filing cabinets, and unidentifiable pieces of pieces, until finally we found the four boxes containing the components that formed my first computer system. As we dragged them into the hallway, cardboard scraping along the concrete floor, I had the sensation one gets seeing something long out of mind but never forgotten. A familiar smell wafted out of one box, and even before my eyes confirmed it I knew what it was. Here was my computer, still wrapped in its old dustcover, a cheap piece of vinyl dyed a walnut brown.

Out of another box came a hinged case made of black translucent plastic and decorated with an elephant head. The day I carefully glued on that sticker with the words *Elephant Memory Systems* was sharp as ever. As I lifted the case it swung open and dozens of floppy disks—large flat black things with donutlike holes in the center—broke loose and fell out, sliding onto the floor. Each was labeled in the blocky writing I recognized as my own: *Starcross, Astrochase, Sector Copier, Basic A+, Assembler, Amodem.*

As I bent down to collect the spill of memories I remembered a long-ago time when I'd carefully packed this same case into my bookbag to take to Roger's house at the outer edge of Queens, an hour's subway ride on a line I'd never traveled before. There we spent an entire Saturday afternoon—two twelve-year-olds pirating software: dual drives churning, cables connected, kilobytes of games traveling over telephone wires, through the modem and onto our disks, with Roger's mom occasionally picking up the phone, bewildered, hearing the warbling screeches and beeps of bits through the receiver and from our room, Roger's wailing "Mom!" and a whispered "fuck" from me. Redial and hope that the pirate board won't be busy, its four 300-baud modems jammed up by kids with speed dialers and a hunger for free software.

In the days when modems cost $600 and had to be hand programmed I'd scored a 1,200-baud modem—four times faster than the usual modems—from my father's office in Manhattan (where he worked building tax shelters). Most of the pirate boards had nothing to match the speed of my modem, which forced me to slow it down to 300. This was the modem I brought to Roger's. His parents, both recent immigrants, were suspicious of what we were doing but couldn't quite identify what it was or why it was wrong; all they could be sure of was that we were maddeningly intent over some nefarious enterprise, a quiet but clearly conspiratorial look about us.

Every time Roger's father came into the tiny room we leaned in toward the small television set, its blue screen and lines of code littered with the verbs PEEK and POKE—commands to examine the contents of the machine's dynamic memory (PEEK) or place a new value in the electric matrix that formed the abstract metaphor of machine memory (POKE). The states of these holes in memory controlled the modem and guided the incoming bits into the magnetic floppy drive, where a double rack of read-write heads were working overtime, grinding back and forth, making sounds like a cartoon kitchen appliance about to explode.

Scattered across the floor were plastic floppies either blank (ready for formatting, the means by which the computer neatly metamorphoses the magnetic sheet at the heart of the floppy into a meaningful array of storage locations for incoming bits) or jammed with fresh copies of the latest games. Roger's father gazed at this chaos in dismay. There was his son, hunched over the computer he'd bought because he thought it would make his son smarter, a better student, a winner in America. Now he wondered if he had done the right thing sending Roger into uncharted waters. Was this what a computer was for, to turn his son into an unruly recluse?

The grown-ups couldn't get this any more than grown-ups of other generations could get rock 'n' roll or tongue piercing or pot. We were doing our thing, hacking, and the rest of the world didn't matter. Anyway, how could parents—or any grown-up for that matter—ever understand something that had not been invented in those ancient times when they were young?

That day I walked out of Roger's house fully loaded, two dozen double-sided disks neatly labeled, and headed for the F train happy as hell. With floppy drives, disks, and modem snugly on my back, I was heading home to my computer to plug everything in and indulge. Here were days, maybe weeks, of new things to explore, a treasure trove courtesy of a nascent phenomenon—the Pirate Bulletin Board: usually one modem, one computer, and another kid with an extra phone line, his computer on twenty-four hours a day and a bank of disk drives ready to deal out the latest cracked software, copy protection broken, any program ripe for infinite duplication. We did it because it could be done, and we were among the few who knew how.

These bulletin boards popped up and disappeared from day to day while others sprung up in their place. Phone numbers passed from friend to friend were posted on other boards with names like Aladdin's, Rainbow, Pat100, Spider's Web, Pirate's Cove. These boards would last until parents figured out what was going on and shut them down, or until their operators got bored or distracted by the fresh call of puberty leading them to glossy magazines with center-folds that told of a next stage of life.

Later that night I called Roger, wanting to know if he'd tried the software: What was good, what was sick, what hadn't been worth our time and trouble? Roger couldn't come to the phone, his father said. When I saw Roger in school on Monday, he told me his father had taken away his computer. It was distracting him from schoolwork. There was more. I couldn't ever come back, his father said. I never did.

As I stood next to my mom it was strange to think of that time after so many years. Holding the computer in my hands. It felt light, its design quaint—with big keys for a child's hands and color-coded buttons. Digital watches and pagers now have more memory, more processing power, than this machine. Yet I'd spent hundreds of nights exploring what seemed a wide world of bits; I'd lived a thousand lives, died a thousand deaths, had been both God and acolyte inside a 2-D world all my own, a 48K universe that exists today in $39 disposable gadgets. How huge that world had seemed! Cheap and omnipresent machines, so easy to slip into; they seduced us—so many of my friends. We just disappeared one day, stepped into the arcades and vanished, reemerging years later as adults.

1

Pong

MY ELECTRONIC SEDUCTION began in 1973 in the bar of a French hotel. The year before, when I was four and a half years old, my family—mom, dad, little sister, and I—had moved from Manhattan to Paris. We stayed five years. Although today I regret none of it, glad to have learned to speak a second language and know another culture, at the time it was a nightmare. To this American boy, the City of Lights was little more than a drab, dark, miserable place. I spent that first year grappling with a new language, starting school, and embarking on what would become a familiar cycle for me—new school, isolation, the first flowering of a friendship or two, then the inevitable breakup as the cycle repeated and I found myself in the next new school.

Close the door, take off, and leave your world behind. We'd done that in 1972. Planes were both fantastic and frightening to me. Fantastic for their size and power and the excitement that came when the machine lifted off the ground. Frightening because they were the engines of separation. One day in October we packed our things, got on a plane, and moved thousands of miles from our home in Manhattan, sent away by a force my dad called "business." My grandfather had died earlier that year, and my dad at thirty-two suddenly and unexpectedly inherited his father's

connections built over years of investments, loans put together to finance projects in faraway countries such as Al-
geria and Iran. My dad, a young investment banker, was
expected to build where his father left off. In my four-year-
old mind this was all inexplicably cruel, beyond understanding. By age twelve I'd already attended five different
schools, having moved from America to France and back
again.

Those first school days in a strange land among the
proudly xenophobic Parisians set me on my course as a professional outsider. In preschool I learned that the world was
divided into two groups: popular children and those on the
fringes. I wanted to be popular, one of the group, liked if
not loved. None of this happened. Instead, I attracted trouble. I was a foreigner with thick glasses, crossed eyes, and
terrible coordination caused by my lack of binocular vision.
On my first day of class I was the butt of jokes that quickly
escalated to blows. In the concrete courtyard of my small
French school during recess I'd accidentally broken another
kid's plastic toy car. Other kids circled around. One stepped
forward, and before I knew what happened I felt a sharp
pain in my cheek. When I touched my face I felt sticky
blood. As I cried and felt the sting of a deep scratch the boys
scattered. When a teacher came by I could feel her disap-
proval; surely I, an unknown kid, was to blame.

I started to spend more and more time in worlds of my
own. I began to daydream, and over months I began to
read. Owing to my double vision I had trouble making out
words on paper; letters moved, words ran together. Reading
felt hard, but slowly, under the guidance of a special reading teacher I saw every day after school, words, sentences,
and eventually whole paragraphs came into focus. One day
when I was seven I found myself lost in a book, the words

making a world so real that I forgot where I was. I'd reached a point where reading had become intensely pleasurable, and I discovered books without pictures; these were science-fiction novels my dad bought for himself, in our family's bookcase.

I read the great classics of space exploration—books by Isaac Asimov, Arthur C. Clarke, Ray Bradbury—and they filled my mind with escapist fantasies as I rode the bus to school or flopped on my bed after school, lying on my side reading, refusing to move until the cramp in a tingling arm or leg became so intense that I had to leave that world and roll over to get the circulation flowing. Here were stories of nomads traveling through deep space, alien ships crash-landing on Earth, lonely astronauts and their families wandering the bleak surface of Mars. Looking back on these books, all written in the 1950s and '60s, I see that what's most striking is the near absence of computers; if they were foretelling the future, they missed an important part of it.

By third grade school got marginally better. As my French improved I no longer made embarrassing mistakes, misusing similar-sounding words. I discovered that among the bigger kids and popular kids were other kids like me. We reached out. One in particular, Jean-Baptiste, was the smallest kid in school. We made an odd couple. I was taller than most; he was tiny. He was nimble and fast in sports. I was not. Yet together we found common escape, swapping French comic books or playing complex games with glass marbles in the paths of a nearby park. When my books or Jean-Baptiste were not around I immersed myself in an addictive game of "let's pretend" that lasted until fifth grade.

The game was simple: I was in fact not human; I'd been sent to Earth to investigate the nature of Homo sapiens. My plan made sense. What better way to learn about this strange species than through the eyes of a child, one who could "grow up," go through the entire cycle of human life, and report back with authoritative inside information? My parents were not my parents at all; they were just the target family selected. Through a highly sophisticated process, the details of which I hadn't exactly worked out, my identity was inserted in vitro into the fetus that came to be named David.

There were great advantages to this identity. I could stand at a clinical distance whenever anyone spoke to me, especially when they were angry. An adult shout became an act to be studied. As angry words cascaded around my ears I took careful notice of the reddening faces, tics, and other physical symptoms displayed by the speakers. The content of their communication was lost, erased. Only the process was observed and preserved. At school, my special mission helped me forget the need to be liked; I could navigate more safely through the halls.

In fact, isolation became a particularly good means of studying the worst in people—sparing me pain as I stored up valuable information for my report to my own race. Every now and then I would question my alien identity, sometimes looking in the mirror after brushing my teeth at night, challenging myself to do something inhuman to prove the story was real.

I knew of a spaceship concealed on the roof of our apartment building to be used only in the gravest emergency. Shaped like a small flying saucer, it had two gull-wing doors that raised themselves to allow entry. Inside were two seats and the simplest of controls. By sitting down I could speak to the ship and navigate with my voice. Inside I imagined a supercomputer like HAL in the film *2001: A Space Odyssey*, an intelligent machine to do my bidding. I could ask it anything, to fly me off to anywhere. I'd fantasize about taking this ship around the world or through time, visiting and exploring other eras or planets.

The French obsession with cinema greatly helped these fantasies. Paris had dozens of revival houses playing old movies. In the 1970s, when French television consisted of two state-owned channels broadcasting nonstop boredom to a nation of millions, children rarely associated the box with entertainment. When I wanted to see something interesting, I went to the movies.

Every Sunday afternoon a dilapidated tiny theater off the Place de L'Odeon in the then-seedy Left Bank offered a celluloid brew of Tom and Jerry cartoons followed by Stanley Kubrick's millennial masterpiece. As a seven-year-old I went as often as my mom and Samantha, my little sister (who was barely four), could stand taking me, certainly more than two dozen times. Seeing that film so regularly became a kind of mystical education for me, the way some children commit to memory verses of the Bible or the Koran.

> "Open the pod bay-doors, HAL."
> "I'm sorry, Dave. I'm afraid I can't do that."

This was the film's exquisite moment for me, when the astronaut named after me faces the full implications of a conscious computer that has knowingly chosen to kill him—the now-familiar threat that an intelligent machine, no longer needing its creator, might decide like Frankenstein's monster to turn on him and his race. Indeed, in the brave new world of computers it seemed that the creation process is not complete until the machine is capable of destroying its creator.

The sparse dialogue and obtuse story line could have bored me, as it bored many adults who saw it; but at age seven I barely heard the words, so drawn was I to the special effects that created a future world, one for which I was better suited than the school yard. Every time I saw the film it reinforced my obsession with outer space and confirmed that one day I would travel there. I was certain that my destiny, and the destiny of all human beings, involved leaving Earth.

My parents tell me that when Neil Armstrong landed on the Moon they woke me up and sat me down in front of the television set; at eighteen months I wasn't interested. Another eighteen months would pass before I knew what the Moon was, what spaceships were, and that people were going up into space all the time, boosted by multistage rockets. I have faint memories of my parents showing me a copy of the *New York Times* with a blurry photograph of a space capsule orbiting the Moon taken by an astronaut through the porthole of his moon lander.

While that picture was an early confirmation that I'd go to space one day, it was *2001* that showed me what my future in space would look like: the space station, the moon base, the long, long Jupiter-bound spaceship with its nuclear-powered engines at one end and bulbous sphere at the other, where the astronauts rested in cryogenic sleep.

In Paris I was comforted by the prospect that such a world would one day exist. Now I know better. When I watch the film these days what strikes me is its loneliness: A crew of two manned the ship, or rather watched as HAL manned the ship, arranging for their food to be processed and their communications sent to Earth. The astronauts were left to jog round and round the white ship or play chess with HAL, who was programmed to lose 50 percent of the time. The environment is desiccated, a machine-mediated universe where each astronaut lives alone, separated from others by layers of technology that stifle direct human contact.

Back then the computer HAL seemed infinitely more interesting than the human protagonists, more interesting even than space itself, that void through which the doomed crew traveled; and so I thought nothing of the loneliness of the astronauts. I probably thought loneliness was a natural part of life. It was for me.

Christmas 1973 we went to the French Alps for a family ski trip in a part of the country that catered to the functionaries, civil servants, and middle management of France. Rarely visited by foreigners, these resort towns had few of the trappings of European luxury. France as a nation has strong socialist tendencies, and these holidays were meant to be communal experiences; all meals and ski lessons were taken together, without class distinctions in either.

We stayed in a huge hotel called Hôtel de France, a wall of windows facing the mountains with a dining room where hundreds ate at long tables, soup was ladled out from communal bowls, and everyone nibbled on the same kind of bread. I loved this part of France because it allowed for safety in anonymity. That holiday in ski school I fumbled on the slopes. I was a first-level novice, which entitled me to a small metal snowflake that I could pin to my ski jacket. Fascinated, I listened as my father explained that when I moved to the next level I would get a pin with two snowflakes, then three, followed by whole new shapes— stars, camels, and for the best skiers, rockets. As I slid awkwardly down the mountain, thinking of becoming the best skier in the world, Samantha took to the slopes with ease on tiny foot-long skis. We fought most of the time, vying for our parent's attention. Because she was younger and the jump from English less jarring, she spoke better French. Born with better eyes, she read and wrote better than I did at her age. At her small school Samantha won a prize for being the best student. She seemed more French than the French. Her success translated into less attention at home: my parents—especially my mother—paid less attention to her and more to me. Every day after school my mother took me to a special reading tutor. My sister was left at home with a baby-sitter.

On our holiday my dad skied with the grown-ups, and my mom—who didn't ski at all—watched us from the valley below. One afternoon I came crashing down the slopes and tumbled into a heap of snow a few feet from where she stood; terrified that I'd broken my leg, my mom called the ski patrol. Basking in all the attention, I decided that accidents were exciting. A few X rays revealed nothing more than a sprained ankle. I was bandaged, given little crutches, and my dreams of two snowflakes and a golden rocket pin were banished to fantasy. Left to wander the hotel during the day and watch people swimming in the bizarre outdoor pool, hot water producing billows of steam in the Alpine air, I discovered something fantastic.

Outside the dining room was a bar decorated in the sparkling, smoky-mirrored chrome that presaged the coming disco era. One afternoon I wandered in. I made my way past the bartender, drawn toward a machine at the far end against the wall. It looked like a television set running a cartoon. I wondered which show was on. As I got closer something seemed strange; I'd never seen a cartoon like this one before. I'd never seen a TV like this before. Where was the channel dial? *What lame cartoon is this?* I wondered, staring at the almost blank monotone screen. I stood watching the "show"—two rectangles batting a square between them—*bonk . . . bonk . . . bonk* went the machine. And then it all became clear. This wasn't a television show; this wasn't a television. It was a machine playing some sort of game with itself! It was showing off, to me. It wanted me to play with it. I grabbed the knobs and spun them around, noticing the coin slot. Back and forth went the blur, jumping across the screen in rapid, barely visible increments, eminently familiar yet totally strange.

Like a Three-Card Monte player, the machine lured me into a familiar game, but one played on its own terms. It invited me to join in. Setting my crutches aside, I balanced on one leg, suddenly coordinated enough to stand without tipping over. Digging through my pockets I found some change my mom had given me for snacks and dropped a coin in the slot. The screen refreshed itself. Holding the knob, I watched as my electronic paddle followed the movement of my hand. *Bonk.* I hit the luminescent ball. *Bonk.* It came back. *Bonk.* Faster now. *Bonk.* Too fast! It shot by. Several rounds later the game was over. I could lose privately. No one to laugh or yell at me for missing.

I found another coin and played another game, my crutches, my silver snowflake, my sister the better skier, school—gone. This was bliss. Here was something I'd been looking for without knowing it. If I had one of these, life would be better; life would be great. There was nowhere else I'd rather be. The sounds of the grown-ups in the hall and the bartender flipping his paper faded away until all that remained was the *bonk bonk bonk* and that silver square traversing the black screen, from paddle to paddle, left to right and right to left, following a clear but barely comprehensible logic. Who or what controlled my opponent's movements? What were the rules governing the flight of that square shuttlecock? Staring at the plastic console with the words AVOID MISSING BALL FOR HIGH SCORE written on it in English, I knew this was going to be a great trip after all.

2
Space Invaders

IN PARIS SCHOOL ENDED EARLY in the afternoon, and sometimes my father would come and take me to his office overlooking the Seine. I'd watch him work, which as far as I
could tell meant speaking on the phone. If anyone asked me
what my dad did for a living, I'd say "talk on the phone."
One afternoon in the fall of 1975 we left his office and
headed out for a meeting. Excited to watch this secret adult
world in action, I eagerly climbed in the car with him. We
drove across town to another building with similar offices
filled with all sorts of toys—photocopiers, typewriters, inter-
coms, paper clips, Dictaphones. Quickly bored by the business conversation between my dad and another man, I left
the room where they were meeting to prowl around the
premises. Adventure and intrigue. At the end of a long hall I
found a huge office with a single desk carved with filigree
and baroque curlicues snaking up and down the legs. The
office was empty. I went inside. A world of treasures: executive booty. Too much to carry out with me. A silver slinky
toy; an electric panel with a mosaic of tiny colored bulbs
flashing in patterns; a flexible bowl made of cardboard with
fluted sides painted rainbow colors holding a glass ball that
rolled around a track; steel balls hanging from strings in a
row, knocking each other back and forth in perpetual mo-

tion. Best of all were the toys on the desk—a heavy gold digital watch with a black face and luminescent red numbers that glowed the color of wine, and a heavy silver pen, thick like a cigar, with a digital calculator embedded in its side, tiny numbered buttons, and a gray liquid crystal screen. This pen, most of all, drew me. I climbed onto the heavy leather chair, a serious seat in the cockpit of high finance; it suddenly tilted back and nearly threw me on the floor. Gingerly I lifted the pen from its leather cup and pressed a button. A number appeared. I pressed another. I was mesmerized. The tiniest calculator ever, at a time when digital calculators were still an extravagant novelty, for the sheik and magnate. I slipped the pen in my pocket and went to find my dad.

I'd stolen a digital treasure at a time when "digital" meant elite, exotic, expensive. For adults, the allure of these $1,000 trinkets was simple: be part of the future first. Out of a child's financial reach, these were toys for grown-ups. I knew the pen was expensive, but for me the thrill was not in the price; pleasure came from its strange flexibility and the excitement of stealing. Unlike other toys a calculator had an infinite number of states: see what the biggest number is, the smallest number. Learn to write words by typing in numbers and flipping the calculator upside down so that 7734 spells HELL and 338 makes BEE. These digital gadgets felt alive and animate, not like an animal but like a place, an environment. A territory existed within that silver cylinder. How mysterious. What were the rules of this place? What were its boundaries? What could be done there? These questions, however inarticulate, impelled me to explore by pressing buttons and looking for patterns. Part of the delight came from a feeling of magic. Open the pen up and all there was to see was a flat piece of green plastic with little metal lines etched on its surface. No moving parts. No obvious purpose to any of these innards, which looked more like modern art than machine. How indeed could pressing tiny buttons perform huge calculations? And to fit it all in my pocket—so private, so portable, yet so powerful. Grown-ups were equally mystified. For a child who always expected adults to understand and explain how things worked it was exhilarating to know that with these digital devices we were all equally amazed. I brought my new toy to school the next day and suddenly became popular.

When we went to a nearby park under the Eiffel Tower to play during recess, the biggest kids didn't bother with the usual soccer game—a game I avoided since my poor coordination was considered a detriment to either side. Instead, they stood around me and waited one by one to use this incredible device that did math in seconds and seemed to prove that school was indeed a place where we learned useless things, a cruel joke meant to keep us occupied during the day. Why bother memorizing multiplication tables when we could all have pens like this in our pockets, one boy said aloud. Back in the classroom, everyone was talking about my magic pen. Even the teachers were flabbergasted. There were no rules governing this sort of thing. Should it be banned? If kids had these at home, would they ever learn math?

The flurry of attention, however, was my downfall. When my mother came to pick me up after school that day in her purple Austin Mini (which resembled an outsized sugar candy on wheels) I was surrounded by more children, showing them my prize. As soon as my mom came out of the car I hid the pen. A child next to me said in French, "Let me see it again." "No," I told him. Sensing a secret, he looked up at my mother and asked, Where had I gotten that pen? What pen? my mother replied, looking at me, her hand reaching for the purloined object. In the car, my face red hot, I turned mute as my mother scolded me and said I would have to send the pen back with a note of apology. I was terrified I'd be punished forever. The letter was accepted, the pen returned. Strangely enough, no one other than my mom thought I'd done anything especially naughty. My father— and even the man whose pen I'd stolen—shrugged it off. The desire for a calculator or a watch rather than candy or a plastic truck signified intelligence, sophistication, a precocious step toward adulthood. These were adult objects and using them hinted at the budding grown-up inside me. In the face of such excellent generation-crossing toys that even adults could hardly resist, the moral standard for little children was lowered. I was forgiven.

I didn't know it then, but what made these new executive toys possible was a remarkable new invention, the microprocessor. Invented in 1969, the chip, as people would come to call it, slowly made its way from America to France. In 1975 my father, returning from business trips to America, would bring back news of strange wonders. Digital gadgets were proliferating in America and prices plummeted as chip factories moved into full production. Within a year the cost of the elite executive digital watch fell from hundreds to dozens of dollars. What had so recently signified adult power and exclusivity now became associated with the antithesis of the global executive, the kid. When my father brought back a digital watch for me, we had a long family discussion about time. Would I never learn how to read a "normal" watch? My parents put an old alarm clock in my room with a metal hammer and two little bells so that I would learn to "tell time." But on my wrist I had a digital watch. Numbers that glowed at night when I pressed a button. I could wave it quickly in front of my face in the dark and follow the patterns from the streaks of light. Those red-faced watches became the first icons of the digital revolution, must-have fashion for millions of children.

In France, a country slow to adopt the latest American and Japanese technology, my access to digital toys was limited. As I mastered French and became accustomed to the culture, America came to exist as an exotic faraway land, a place from where the future came. The faint echo of change, however, would soon become an exhilarating blastwave as my parents announced in 1977 that we'd return to New York, a city that, even though I'd spent five years in Paris, I still considered my true home. At school, when the kids asked me, "Tu est américain?" I always answered back, "Non, je suis New Yorkais."

When we returned in the fall of 1977, the streets, the buildings, the shops of New York were alive with a kind of giddy chaos. Where Paris had been clean and well organized, flowing along stately rhythms, here in our new neighborhood bordering the northern end of Midtown and the southern end of the Upper East Side was a constant swirl of noise and color. Graffiti lined the side streets; crowds of people came together and broke apart on Lexington Avenue, where two giants, Bloomingdales and Alexander's, anchored a strip of electronics shops whose window displays contained ever more exotic and sophisticated technologies all jostling for space: Walkmans, VCRs, digital watch-calculators, digital alarm clock radios; bright lights, mirrors, neon calling out to New Yorkers and tourists who came from all over the world to stock up on the latest magic from the high-tech labs of Japan and California.

I had arrived, swept up by that modern magic carpet, the Boeing 747, at the epicenter of paradise. Rich with sugar cereals, video games, electronic toys, cartoons, and science fiction, this was a place where television had twenty or more channels and children's programs were saturated with ads for toys. Everything was bright and wonderfully gross: explosive Pop Rocks that sizzled on your tongue, gooey Slime that came in plastic garbage cans and glowed in the dark, collectible cards featuring zits, excrement, and pus. Sandwiched between cartoons, the latest toys were paraded before us for approval or rejection in the giant amorphous arena formed by broadcast signals. The TV became my companion and first friend in New York.

Soon after we arrived my parents announced their separation, a piece of news my sister and I greeted with feigned nonchalance. Our father moved to a hotel five blocks away. The one television in our apartment, quarantined in my mother's room to limit our access, became my primary destination. On Saturday mornings my sister and I would stand outside our mom's door, debating whether it was too early to barge in, wake her up, and turn on the set. With our dad out of the house we became much more unruly, a tempest of kinetic brattiness whirling around my mom, who had to contend with our post-separation feelings as well as her own.

Whatever the reason—the cartoons, color TV, games, and sugar cereals, or the end of having married parents—our hyperactivity became an emotional battering ram. At the incredibly late hour of 8:30 A.M. on Saturdays, it was time. We'd pound our way into my mom's room, flip on the set, and romp around her bed, my sister and I pinching and slapping each other silly, dueling over the dial, two demographics in perpetual enmity—I, the Boy, revolted by My Little Pony ads—she, the Girl, disdainful of Transformers and Big Trak. For me, Big Trak was the sine qua non of the new electronic toys. It caught my attention one Saturday morning in early 1978 during a lull in the fighting between my sister and me. Up came the ad: Big Trak was big—four chunky black wheels and a rugged, gray truck tearing through the halls of a big TV-land house. It shot down a hallway, camera at floor level. The family dog jumped back, the parents looked dazed and foolish. The kid was in control.

You could tell Big Trak where to go by pressing buttons on a keyboard built into the top of the truck. The buttons told it how far to move forward, turn left or right, or reverse. Big Trak remembered what you pressed. This wasn't just some dumb remote-control truck, I thought. It had a brain. You could teach it to follow lots of paths. *I want one,* I said to myself. *I want a toy like that. I want to make it run around my house.* Everyone in the ad, except the kid, was transfixed, thinking that magic made the machine dart behind corners. But the kid knew better. He knew he programmed Big Trak. He had a secret, a strength of his own. That truck was in fact an ambulating computer, a digital machine that came into the home through toys, through a child's world. The implicit message behind the ad spoke of seduction: power for little kids, a power grown-ups couldn't grasp.

Grown-ups had sailed to America and landed on the Moon, but in this new territory the pioneers would be kids. We could fulfill the ancient childhood fantasy of bettering our parents. Such dreams had always existed in children's books and movies, bounded by secret gardens, pirates, and magical kingdoms. With digital toys the dream became real. Big Trak was a harbinger of the tectonic shifts that soon would take place, changing the meaning of childhood. Children soon would be masters of something adults didn't understand.

Two events in 1977 mark the beginning of the Digital Age. In April the first mass-produced home computer, the Apple II, was released to instant acclaim, and in May *Star Wars* opened. The film introduced our generation to the magic of special effects—plausible alien worlds, real lasers, hyperspace—and its success stimulated a desire for toys that were technologically advanced and futuristic. *Star Wars* and the Apple II inspired the end of toys as we knew them as companies quickly positioned themselves to gain from popular culture's fascination with technology. Using microprocessors they accelerated production of a few experimental gadgets, sensing a potential Christmastime boom.

As the 1977 Christmas toy season gained momentum toy analysts noticed a strange trend—the two hot items of the fall season were both electronic. Merlin, a handheld computer game that resembled a red telephone with eleven circular buttons arranged in a grid that flashed lights and emitted tones (depending on what you did), was developed by Parker Brothers. Simon, by Milton Bradley, asked you to tap out patterns of sound and light on four large buttons embedded on a black plastic disk the size of a plate in a game loosely based on Simon Says. Both were selling out in stores across America. By the end of the year $21 million in electronic toys would make their way into people's homes, up from practically zero the year before. News stories described despondent parents lined up outside toy stores, desperately seeking Merlin and Simon. Right behind Simon and Merlin in popularity were a series of handheld games made by Mattel—Battleship, Football, Auto Race and Comp IV. Each sold for under $40 and offered the irresistable experience of sitting alone and playing a game with a wily opponent. In 1978 electronic toy sales grew to $112 million, in 1979 to $500 million. In the end Merlin outsold them all, grossing $120 million by 1982; at the time it was the most popular game ever made.

Designed to be small and lightning-fast, these digital toys adjusted their reaction time to your skill level, thereby creating the illusion of intelligence. Where kids once played with miniature replicas of adult artifacts—cars and tea sets, playhouses and paper dolls—toys now were unrecognizable. Instead of trucks we had plastic car-racing games in which the "cars" were flashing-red diodes, slices of light that symbolized movement. Throughout history toys prepared children for the future by allowing them to imitate adults. Children fought fires, built houses, sailed ships, saved lives. Toys reflected the values of the time. Our toys were different. For what future were they preparing us? What role were they teaching us to play? Ours was the first generation to have toys that bore little relation to the world of adults, or reality on this planet. They came from a galaxy far, far away, as the opening titles to *Star Wars* suggested. My future was not here, it was somewhere out *there*, and my toys were all about escape, leaving Earth for whatever existed in space. *Star Wars* and video games made the fire engines, miniature horses, and plastic soldiers a bore.

The chip makers worked overtime to supply our craving for the grand illusion. General Instrument, a microprocessor manufacturer, powered many of these toys. By 1978 they'd sold 7 million of their AY38500 chips at approximately $5 apiece, turning the toy industry into an electronics industry. In 1979 when the Japanese company Taito released an arcade game called Space Invaders, the Japanese treasury ran low and began minting more coins. When the game came to America, a similar frenzy swept across the country—sales of video games that year increased by 300 percent. Our minds were ready to embrace the unforeseen. As waves of "space invaders" descended like a Greek phalanx in formation, your laser cannon darted between bunkers on the bottom of the screen and fired into the creatures until you killed them all and another, faster, wave arrived. With Space Invaders the familiar video games—Pong, car racing, bowling—seemed as silly as the toy horses and plastic cars. Why play virtual tennis when you could blast invaders in deep space? Now the future seemed limitless. In an escalating arms race of the imagination new electronic worlds were invented, explored, and discarded, each iteration upping the ante of verisimilitude and exoticism. Toy makers, computer companies, and filmmakers were happy to oblige.

Part of the success of digital toys lay in their nature: they were not just toys, but *playmates*. I never realized how lame my toys were until I saw Merlin. My old toys—Star Trek dolls, plastic guns—were dead things. They didn't move the way these new toys moved, and they weren't smart. These inventions represented a new, digital way to play. They played back, reacting to what you did. And in this was something entirely new: an experience of communication—dialogue—with an inanimate object. We actually played *together*. This would become the singular characteristic of a digital world—machines that felt alive, malleable, responsive, changeable over time. Here was the sense of finding extra life, of going beyond ourselves into another world, of being an explorer on the outer edge of a place that before never existed.

When I came back to America nine years old and friendless, digital toys were my first playmates. I didn't know what to expect from the children in my new school. The Fleming School, two blocks away from our apartment, offered a bilingual, French-American education. Here I could preserve my French language skills. The school was tiny—a slender, five-story townhouse—but to me it looked vast, imposing, and filled with uncertainty. Children from fourth through eighth grade co-existed in this white-stoned building with cramped classrooms, narrow halls, and a single elevator in the center (for sick students and teachers only). Samantha was in the lower school (kindergarten through third grade) in another townhouse down the street.

I arrived at Fleming feeling nervous and dizzy. It was a torpid, humid September day. I held my mother's hand, but as soon as we got within sight of school I dropped it. I didn't want to seem afraid or weak. I was acutely aware of having come from another country and worried that I would be ostracized by the other kids. At the school's entrance I hurriedly said good-bye to my mom. I hoped to get in quickly and hunker down, safely investigate the lay of the land, and find my place. I did not want to stand out. I wanted to blend.

As soon as I entered the third-floor classroom a wave of giggles greeted me. I knew something was wrong. I looked down at my gray-flannel shorts and tucked-in white shirt—the typical outfit of a French schoolboy. "You can't wear shorts!" one kid said, sneering. "Yeah," another chimed in, "*Stupid!*" I sat in my chair quietly seething, furious at my mother. She should have known the school had a dress code: blazer, white shirt, and either gray or navy-blue long pants. *Why did she do this to me?* I wondered, scratching at my desk with my fingers.

My mother—who'd always paid so much attention to me—had become less attentive, and I craved the old ways before the separation. At home she would cry for no reason or sleep later than usual. I felt as rootless in school as I did in our new fatherless apartment. Antically wandering the halls, tapping against the walls, I found a way out by daydreaming more intensely, reading, and playing electronic games. Being in my room had its advantages—I could avoid my sister and my mother, and hide from the experience of seeing once-powerful adults falter. A few weeks after that first day of school I came home to find my mother in the living room speaking quietly and sobbing softly into the telephone. Afraid to go in, I went into her bedroom and stood in front of the silent television set, the faint hum of traffic outside filling my head, and gently lifted the receiver. I listened as my mother, speaking to another woman, said things she would never say in front of me or Samantha. I closed my mouth tight and tried to breathe quietly so they wouldn't hear me.

"I have nothing, I have nothing," I heard her say over and over between tears, and the woman on the other end said of course not and tried to help my mom, but all she would say back was "I have nothing, I have nothing." Suddenly, as if my life depended on it, I wanted to make everything right for her. To take care of her the way she had for me. I put down the phone after she did and ran down the hall, the walls wobbling through my tears. I ran into the living room and up to where my mother sat and put my arms around her and said I love you I love you and I will never leave you. We cried like that together for a while and then, the opening inside me closing, I went back to my room, to my books and Merlin and Simon and my handheld racing game and slipped away to a place where no one could see me and the world was safe.

Simon and Merlin offered instant teleportation to fantasy land. Flip on the switch and Merlin dealt a hand of black-jack, played tic-tac-toe, or made forty-three musical notes with scores of tunes to enter in and play back. Simon never got tired. Inside a chip drummed out sequences of color and sound, colored buttons lighting up, forming a pattern to re-peat Simon Says. My Mattel games simulated football, car racing, and submarine hunting. Each used thin red light-emitting diodes (LEDs) to produce patterns behind a plastic screen with painted overlays. The car-racing game divided the screen into vertical lanes. My job was to accelerate a red LED, "driving" between oncoming flashes of light that symbolized rival cars. The better I moved the faster they came. Playing back at you. The submarine game Battleship let me *ping* my sonar at shadowy subs below, and then, if I was right, I could dump explosive depth charges over the right coordinates. A direct hit sank the little LED-sub. I would sit in class with the game wedged between my thighs with the sound off, twiddling, twiddling.

Although I loved books, escape through electronic games differed from escape through reading. Both created environments that required an active imagination. For the literate, words on a page formed an illusion. To the uninitiated, pixels or diodes on a screen were as abstract as letters are to the illiterate; however, they too served as the foundations of immersive environments. For the multiliterate able to read and dexterous at video games, the difference between the worlds of books and games was substantial. Books offered information and realism that my handheld games couldn't. They offered a connection to grown-ups and the world that games ignored. When I read anything—from my monthly issue of *World* magazine published by National Geographic to the *Foundation* trilogy by Isaac Asimov or my American history book with a chapter on the 1960s that taught me about Martin Luther King, Jr. and the Vietnam War—I was in touch with a fragment of human knowledge. Each book was part of a massive whole made up of everything ever written. Reading offered a passport into an adult world. The power of reading was a power adults could understand because they too were readers. With digital games, however, I was at the threshold of a new technology and new form of knowledge transmission. It was like being alive at the moment the first writers appeared. Where books were static and unchangeable, games were fluid and dynamic. Clutching a handheld game with a strip of flashing diodes, I willed the lines and dots into being, and together we interacted. Of all the old media television required the least participation; you willed nothing. You merely sat back and watched. Adults couldn't share in the excitement or understanding because (with few exceptions) they did not genuinely want to sit down with a small plastic game and play for hours, for weeks and months and years. Neither did girls.

Girls at school rarely showed interest in these new games. By fourth grade we'd begun to be fiercely segregated along sex lines. Boys and girls stood apart, and crossing over was impossible. Like adults, girls were not invited to play with us. This new world of video games quickly became seen in the media, at school, and in homes as a boy thing. After the initial shock of my first day at Fleming, electronic games became a way of making friends. Equally new to all, we were all novices; there was no history here for me to catch up with. Because we played in groups and converted new kids by sharing toys, I'd sometimes pass along my Mattel games during recess or in the morning as we stood outside, waiting for school to open. Learning the games became part of socializing. You didn't merely stay at home, shut in and gaming all the time—you boasted about your score, showed off your tricks, and learned from others who knew more. While the immediate interaction between player and game was solitary, the greater interaction was communal. The first generation of games drew on competition and conflict for drama: kill or be killed; be first or last. The physical gestures needed to play—squeezing, tapping, jerking, pushing, pulling—were aggressive. Where books were quiet, games were noisy. Instead of flipping a page I'd be twiddling and grunting and cursing, shaking the little plastic box.

My newfound interest in electronic games reduced time spent watching television. Apart from Saturday mornings, I watched less and less TV. My sister, however, kept watching. Unsupervised, she would spend hours flipping through cable stations, her seven-year-old brain consuming all kinds of programming, much of it chaotic and adult. Years later she admitted that much of her inspiration to become a "bad girl" came from TV movies, where kids who grew up fast were a favorite topic. The 1970s, with its now famous "latchkey children," was a time when demand for "true stories" about runaway kids, adolescent prostitutes, pubescent hustlers, and dope dealers broke what now seem quaint notions of television propriety. When my mother went out, my sister and the baby-sitter would lie on my mom's bed glued to the set until it was time to sleep.

I collected as many electronic games as I could. Proximity helped. A few blocks from my new home was my Mecca—FAO Schwarz, the ultimate toy store. There the latest electronic games arrived, and after school or on weekends I went, a lone fourth-grader entranced, lusting after every possible electronic game. As 1977 rolled into 1978, games appeared with ever-increasing frequency. Where once Christmas and gift-giving grandparents defined toy seasons, these electronic friends defied all limits. People bought them year-round. They collected them like dolls in series.

One afternoon in 1978 I went to FAO Schwarz to check out the toys with my new friend, Jesse, a wiry olive-skinned kid from Fleming, who like me had come from far away, from California. A great big carpeted stairway led up to the second floor, where all the good toys for boys were (downstairs were fluffy toys for girls). Ascending those stairs that day, Jesse and I came upon Big Trak. On a pedestal. It was the first time I'd seen it in real life. Untouchable, it sat motionless inside a plastic cube, like the crown jewels. We stared and fantasized out loud.

"I'd have it carry stuff around the house," Jesse said.

"Put firecrackers on the top and see what happens," I offered.

We talked for a long while, dreaming of Big Trak and new variations.

"Yeah, one that goes underwater and can shoot things," I suggested.

"A demolition derby where they crash into each other," Jesse added.

I'd be ten years old in a few weeks. My first two-digit birthday. Perhaps on this occasion my mother would get me Big Trak, if I asked.

When the day came my mom handed me a big wrapped box with FAO Schwarz stickers on it. I tore the wrapping apart, ignoring my grandma's advice to save the nice paper, sitting on our living room floor, grinning. The size was right. Big red letters on the box revealed what I'd hoped for.

For days afterward I sat in class plotting paths for Big Trak to follow. I'd measured my apartment, including the space taken up by furniture, and made diagrams on graph paper that I hid in my notebook. While the teacher wrote on the board I drew neat little lines from one point to the next—designing convoluted routes between couches, tables, and bathrooms. Since our apartment had lots of furniture my plans favored sinewy, knotted paths with few straightaways. The instruction manual said that Big Trak could move up to half a mile. That would be ten Manhattan blocks. Ten blocks! I pictured Big Trak trundling along Park Avenue intimidating dogs and flustering doormen. I thought about putting a booby trap on it—raw eggs or a water balloon—that would break and splatter people, but the truck seemed too fragile. A puddle, let alone a curb, would derail it from the course I set. So I satisfied my grandiose fantasies with at-home experiments, putting obstacles in its way, ramps and bridges to cross. One afternoon I brought Big Trak to Jesse's.

Jesse was my best friend; we had a lot in common. Our parents were split up and we didn't have many kids to play with in a new city. Jesse lived across the street and we alternated between apartments, although his made a far better playing field for our games. His mom was into a Japanese religion she called Zen. Every surface in that apartment was white—white wall-to-wall rugs, white walls, white furniture. I decided that lots of empty white space was part of Zen. Jesse's mom lived in fear of spills. She forbade us to take our favorite chocolate drink, Nestle's Quik, out of the kitchen. We loved the white rugs because they offered a blank screen for our imaginations. The apartment served as one enormous 3-D space where our high-tech fantasies from *Star Wars* and electronic games could play themselves out. Dozens of metallic action figures, board games, anything we wanted to build could be placed on big open surfaces. Big Trak could roam free here. One afternoon we had a marathon play session, meticulously measuring everything, and built an extreme obstacle course with slanted ramps of books, jumps, bridges, and a tunnel. Jesse's mother had a Siamese cat, and we tried to get it to hunt Big Trak by sticking cat food on the roof. It didn't work—the food fell off when Big Trak rolled over a book, which left a brown stain in the rug.

Big Trak was the first computer I ever programmed. That's what I was doing by pressing the keys on its roof. Beginning to relate to a game as programmable was a leap in abstraction. Here was a form of responsibility, of active participation, thinking, and analysis that crept into my time with Big Trak. The process was instinctively modular, a breaking apart of goals into subgoals, building back up to the whole from the smallest unit of problem solving. The act of laying out graph paper, modeling a room, and associating each square with a unit of distance meant I had to measure the room first and then think about what scale to use. Each square served as the smallest unit of measurement and gained meaning by pulling back, much as dots in a newspaper photograph or television screen fuse together when looked at from a suitable distance. I used a lot of math to make Big Trak work. At school I consistently received C's in math, yet at home I eagerly applied principles of arithmetic and geometry. What made these laborious tasks worthwhile was the experience of making a finished product that happened to be thrilling to a ten-year-old.

Toy companies understood our needs and built pieces that when infused with proper narrative, vivify themselves into a living environment. Toys come alive the way everything comes alive—by changing over time; the accretive nature of time turns a piece of plastic into a character. Putting time into toys is what the first electronic games did—those red diodes zipping past my race car had a beginning, middle, and end. Some shrewd game companies realized they could put time into toys *without* electronics by harnessing our willingness to suspend disbelief and infuse bits of plastic—even paper—with the illusion of change over time. This is an old strategy in board games, and in the mid-1970s it led to a new, fantasy role-playing game. The classic of the genre was Dungeons and Dragons.

D&D, as we called it, was a metaphor for the thought process behind computer programming and a highly social activity. Through its emphasis on modeling (in this case,

modeling a fantasy world) and requirement of social interaction, D&D foreshadowed much of what I would come to
see as the most important aspects of computers as a collaborative tool for thinking and creativity. It was an example of
how a subculture built by kids would work its way upward
into the cultural mainstream.

Jesse introduced me to D&D. With his old friends in California, Jesse had been a "Dungeon Master"—the person responsible for running the game. The Dungeon Master was
god of that world, acting as referee and narrator. Jesse
showed me a box full of D&D paraphernalia, which he kept
in his closet. Inside were notebooks with carefully written
tables of numbers, maps, and notes. Jesse withdrew plastic
bags filled with many-sided dice—five sides, twelve sides,
twenty sides—and hardcover books with illustrations of
monsters, treasures, and magic spells. That afternoon we
decided to start our own D&D game at Fleming. Jesse would
be Dungeon Master. First, however, we needed to recruit
several more boys.

At Fleming were two other boys in our class of sixty with
whom I'd become friends. Like Jesse and me they were new
kids, and not too good at sports. One of them was partly
deaf. Brian wore a hearing aid in class, and people laughed
at him because he spoke in a loud, nasal voice. The other
boy, Doug, was a recluse. He hung back during gym class
and rarely spoke to anyone. He was, however, an excellent
student. Both Brian and Doug agreed to a play date on a
Saturday afternoon where we would try out D&D at Jesse's.

The game was a great success. It began when Jesse's
mom prepared a big stack of peanut butter and jelly sand-
wiches with chocolate Quik for all of us and ended five
hours later when Brian's mom came to pick him up. We'd
quested through a medieval land that resembled J. R. R.

Tolkien's *Lord of the Rings*, a book we'd all read with great admiration. Like us, sprawled on Jesse's floor, the players in Tolkien's fantasyland sought treasure, solved puzzles, slayed monsters, and gained wisdom, experience, and strength over time. Weekend after weekend we would meet, until our characters and world felt like an extra life, as real as our own in school and at home. Each session was a continuation of the same game, and as time passed our investment in this world became enormous. It felt eerily real, and we began to prize our "experience points," which were the most important benchmark of our character's evolution. I was a magician, Doug a thief, and Brian a warrior. Together we formed a band of travelers. Jesse created the monsters and obstacles that crossed our paths.

Helping Jesse were books sold by the company that invented Dungeons and Dragons, Tactical Studies Rules (TSR), based in Lake Geneva, Wisconsin. Jesse studied the books and took notes on how various combinations of attributes such as charisma, dexterity, intelligence, and strength might combine to form different characters. He told a good story, and kept good rules. A bookshop in lower Manhattan on a grimy industrial block filled with textile wholesalers specialized in fantasy games, and they sold every TSR book, special maps, and all kinds of dice. From time to time our parents would take us downtown on a D&D field trip to restock on fantasy stimulants.

We bought every book. Page after page of illustrations of wonderful monsters detailed their strengths and weaknesses in absolute numbers, along with bits of story explaining the genesis of each beast. Other pages contained lists of magic spells, along with explanations of their powers. Some spells required scrolls; others used potions, staffs, or simply words. Pages of weapons—swords, battle-axes, maces—were followed by lists of armor. At the age of ten very little effort was required to slip into this world.

Creating and nurturing a D&D character became an obsession. By night I read Tolkien and weekends I found myself—along with Jesse, Doug, and Brian—entering woods filled with elves, questing for treasure or magic potions, forming alliances, and sometimes betraying them. For some adults, however, the game had a sinister air. As D&D became more popular religious groups denounced it as a cover for Satanism and paganism; others compared it to drugs, a dangerous addiction, citing reports that one teenager despondent at the loss of a long-nurtured character in a battle had jumped out a window, ending his life. In a decade when juvenile truancy had reached an all-time high the irony of chastising kids for sitting together and fantasizing was not lost on our parents. They thought the game was harmless and far safer than letting us play in the park or the streets.

Tactical Studies Rules provided us with the vocabulary, or commands, that made the game possible. They published the books with lists of all the objects that might comprise a story. The numerical values assigned to each object were cleverly thought out by TSR, designed to support our imaginations. Without them ten-year-old Dungeon Masters would have had to acquire a whole new set of skills in mathematics and statistics—skills that in fourth grade were hard to come by. For instance, a broadsword might wound fatally but be so difficult to wield that the odds of striking a blow were low, unless the character was experienced and of a certain species. Men could wield these swords better than elves (who were smaller). Conversely, elves were better archers, but bows are useless in close combat. Multiply all these objects against each other and the game became complex, engrossing, and very much about strategy.

Together we acted as computers, stringing the objects together (sword-meets-magic spell) and computing their relative values and outcomes, dice tumbling on the floor, pencils in hand, clutching paper listing our possessions and current state of health that when calculated and recalculated led in turn to a new cycle. The games we played began to alter my abilities. Up to then my analytic activities were limited to theoretical exercises in math or science class, like seeing what happened to plants when we stuck them in a closet with no light (they turned white and drooped, or pointed to the seam of the door if any light came through). Now, of our own free will, out of school, we were taking on problems—math, probability, mapping—the mechanics of which were rarely called upon for most ten-year-olds. More subtly still, we were doing a special kind of problem solving, what some might call systems analysis.

In D&D we were creating a large, complex system with history, time, and a future. Multiple "inputs" affected the outcomes of any turn in the game. Figuring out what was best for you and your team and balancing that against uncertainty required the ability to see both close up and far away at once. We were seeing individual objects as part of larger objects that in turn formed a complex whole. The same kind of thinking would be reinforced later as we discovered computers, which came as both mechanical systems (connections of printers, disks, monitors, and central processing unit) and virtual systems (connections of separate programs within an operating system). Some of us would network our home computers into still more complex systems of connected computers. The social space of D&D would later reappear in shared programming experiments and online electronic bulletin boards.

Our little group didn't survive the summer. I went away to camp, played D&D with other kids there, and came back to New York feeling comfortable at Fleming. In fifth grade, however, social drama at school began to hold my interest, especially flirting, and I started to make friends with some of the popular boys, the ones who had intimidated me the year before. Two of them, blond-haired brothers, Eric and Tim, became my closest friends, closer than Jesse. I dreamt of being like them, strong and fearless, with freckled white skin and blue eyes. They would jump from high places in Central Park and run across the street when cars were coming. They would steal toys from stores; Eric had even kissed a girl. Compared to these adventures, earning experience points at D&D seemed wimpy and childish.

Fifth grade was a membrane separating our prepubescent selves from the changes in our bodies. As the year began, one by one some of us discovered stains in our beds—red ones for the girls, clear ones for the boys. Our games started changing, too. At birthday parties (one of the few times boys and girls played together outside of school) kissing games emerged—Seven Seconds in Heaven, Spin the Bottle, Truth or Dare. At one party I caught up with Eric by kissing a girl named Allison on the cheek. From then on she and I would pass notes to each other in class with little hearts on them.

Our discovery of flirtation, of innuendo and intrigue, created a new game, one compounded by the crescendo of the sexual revolution, which crossed traditional barriers to kids. New York's Channel J—the "dirty channel"—gave us strippers at night when our parents were out; *Playboy*, *Penthouse*, and *Hustler* offered stories of strange "swing parties," sort of like our spin-the-bottle games but deeper; orgiastic discos such as Studio 54 captured popular imagination, buffeted by the songs of Donna Summer and Diana

Ross. We watched and learned and began to understand that sex, drugs, and disco were what it meant to be grown-up. We could grow up faster, be older—which was better—by aping the grown-ups. Some parents obliged our fantasies. One girl had her eleventh birthday party on a Saturday afternoon at a club called Xenon, where we discovered a second floor filled wall-to-wall with big slippery couches, like a giant mattress. Others had birthday parties at the Roxy, a roller-disco downtown, where we could groove all afternoon, spinning in circles to "Bad Girls" and "Knock Knock Knock on Wood."

As my parents negotiated the intricacies of their separation, moving toward divorce, I discovered that being with friends was the best way to make myself feel better. And most of my friends seemed to be living with only one parent anyway. We formed hazy packs of shifting alliances. In the final fading years of disco, as we entered sixth grade in the fall of 1979, the monotonous *thunk thunk thunk* of Space Invaders made its way into the Roxy, a flashing-light machine wedged in near the entrance. Space Invaders changed everything. Whereas five years earlier Pong had marked the beginning of video games, Space Invaders signaled a new phenomenon, an activity we could claim as our own: the video arcade with the gunslinging preteen desperado.

At the Roxy on weekend afternoons, while girls spun around the rink in bright headbands and sequined roller-wheels thrilling to the music from *Grease*, I stacked my quarters along the edge of the Space Invaders machine. Space Invaders had spread, infecting every neighborhood, deli, and candy store—any place that sold anything to kids. They displaced the pinball machines. Like pinball, the arcade games offered every neighborhood kid a chance to be a hero, a winner, a pro; but where pinball was analog—limited by Earthbound laws of physics—these digital games had no apparent limitations, no physical constraints on what could be imagined and brought to life. Arcades became clubhouses. A fetid sandwich shop in Midtown with a cigarette vending machine offered me hours of Space Invaders and Marlboros steps from my home. Being good at video games made you feel tough, like you ought to smoke and push little kids around; in just a few months I began to morph under the hothouse energy of pixels. Suddenly I was a smoker. All at once I was a bad boy.

Eric, Tim, and I would layer our quarters along the warped plastic console pitted by a legacy of smoldering cigarettes forgotten in the excesses of a particularly challenging session. Space Invaders asked very little and gave simple pleasure in return: two buttons below the left hand to move my ship from side to side between the bunkers separating it from row after row of space invaders; a round button under my right hand for shooting bolts of electric laser lightning. As the rows came closer to the ground and the cannon brought individual aliens down, tearing holes in their rigid formation, the invaders moved faster and faster until either I killed the last one or they crushed me. Then another, more powerful wave would come moving to the relentless rhythm of the game's sound effects, *thunk thunk thunk,* until finally, inevitably, I would lose. There was no way to win at Space Invaders or any of the arcade games that followed. Instead, we fought for more ships, extra lives earned by scoring high.

I played not to win but to experience the environment, to escape. All early video arcade games offered the same destiny: losing as the ultimate fate. Space Invaders worked because it created a particularly seductive environment, a way to go to another place and slip into a trance, where all that exists is the pixelating screen marred every now and again by the reflection of faces. Eric, Tim, and I would play for hours, feeding quarter after quarter, smoking cigarette after cigarette, and still they would come, and everything would be very much the same until it would be time to leave and outside the day had turned into night.

3

Experience Points

THE EASTERN EDGE OF MANHATTAN near 86th Street contains a sliver of green called Carl Shurz Park. It's built partly over the East River, covering both the shoreline and the FDR Drive, which briefly runs underneath and reemerges around 80th street as an outdoor roadway. This park is relatively inaccessible to anyone who doesn't live in the neighborhood. The nearest subway is five long blocks away. Once you're there, though, the park offers a fine view of Queens and several bridges, a few playgrounds, and the mayor's residence, Gracie Mansion, which sits in the middle of the park. On most days the grounds are shared by groups of elderly people, who live in nearby high-rises and brownstones, quietly sitting on benches along the esplanade facing the river, and schoolchildren out for recess with their teachers. It was here—lying on the side of a small hill early one August evening in 1980, the sky flat and gray, the wind from the river colder than usual for summer—where I first smoked dope.

The smell dissipated quickly. No one walking along the nearby paths seemed to notice three preteen kids passing a brown wooden pipe from one to the other. The park was good for that. No one was likely to surprise you. Tim and Eric lived nearby in a white-bricked apartment house abutting the park's southern end. If it weren't for this proximity the park would have made an unlikely hangout. The three of us were together again, after a long summer. I'd just returned from camp a few days earlier, and when I called Tim he told me that he and his brother had tried marijuana. I was curious.

The real thing was fun all right. As the day slipped into evening I smoked from the pipe until my body became light, and it was hard to breathe or smoke more because we were laughing too much, rolling on the grassy hill, and the dizziness and the high and everything else felt so good that nothing else was there anymore except our secret, ourselves together, friends. When we left the park there was no more to smoke. It was dark and the street lights along East End Avenue seemed closer than usual, brighter; the cars were softer, like pillows. I headed home in a changed city, past familiar things now unfamiliar, nothing quite the same.

As I entered seventh grade my interest in toys and games and computers waned, replaced by a rush to get older. Once again I was a new kid in a new school—the fifth school in eight years. My new school, Horace Mann, wasn't in another country, just in Riverdale, the northernmost part of the Bronx, forty-five minutes by school bus. I was old now—twelve years old—and this time I didn't blame my parents for the change. Before, when we'd moved to France and back again, switching schools and countries had been their fault; since Fleming ended in eighth grade, however, leaving at the end of sixth seemed reasonable enough. My dad explained that switching now was best because with each passing year getting into a good school became more and more difficult. I had mediocre grades and my chances would be better if I left Fleming now, before I slid down any further.

As always, being a new kid wasn't easy, but this time I had a way out. Fitting in didn't matter. I was on a mission of my own. Becoming older was all I cared about, older than twelve, older than a child. And the best way to get older was to collect grown-up experience points. Seventh grade was when I understood that my life would be crystallizing into a pattern that would affect my adulthood; it was a year when I felt that this time the game was real. More real than Space Invaders, Big Trak, or Dungeons and Dragons. More real than the fantasies Jesse and I made up on his expansive white rug.

As fall progressed and I navigated the cliques at my new school, I continued to hang out with Eric and Tim. We were joined by two other boys: Tony and Matt. Tony had gone to Fleming with us. He wore thick black glasses and army fatigues too small for his pudgy body. He was into math and science. Matt was a new friend whom I'd met at summer camp. He was tall and lanky and had thick, black curly hair. He lived downtown with his mom on Bleecker Street in a small railroad apartment. Matt loved to fight, and he'd been thrown out of camp for hitting kids; if I was too high, disoriented by the bars on Bleecker Street or near Little Italy, I could count on Matt to get us out of any situation without abandoning me. Once we'd eaten a bagful of raw hashish stolen from Matt's mother and found ourselves craving hot dogs. At a hot dog stand we ordered six between us, and when it came time to pay neither of us had the money. Matt told the vendor he was sorry and that we'd be back; he promised we would get the three dollars. Matt made the vendor laugh, and a few hours later, when we were no longer tripping over each other, Matt and I returned with money from the bookbags we'd dropped off at his house. There was no one I wanted to be with more when I was too high.

That season the five of us thought we were ascending a mountain of cool. Every weekend we would come up with new tests of who could go farthest: first pot, then a pungent fluid called Rush, glue, nitrous oxide, booze, and downers; firecrackers, M-80s, blockbusters. Then shoplifting, and later (for a few), sex, mugging, and prison. We were climbing, for sure, but we were headed down, not up. No more experience points on paper, no more Dungeons and Dragons sitting at home. The magic potions, weapons, and adventures were to be had outside, in the streets. Points were gained by taking chances that others wouldn't. Doing drugs gained points; stealing gained points; blowing things up gained points; going further with girls gained points. Once someone did something new, everyone else was expected to follow—this formed the next level and the foundation for the level after that. As we climbed through, however, the outside world took our games more seriously. We were starting to do things with consequences larger than we grasped.

The southeast end of Central Park looking down the Plaza Hotel and the posh shops of Fifth Avenue was a favorite hangout, a place to buy bags of marijuana from the dealers who sat all day on the park benches gazing at hotel doormen. On most weekends I would go there with Eric, Tim, Tony, and Matt to buy nickel and dime bags. Back when a nickel bag yielded five fat joints, we'd smoke up on the rocks, watching tourists riding in horse-drawn carriages that left a stench of dung in the air.

At home I concealed everything I could from my mom and sister, preferring to sit in front of the TV playing video games on my Bally Astrocade. My sister's room was next to mine, and she seemed oblivious to my slide. In fourth grade she was into girl things: horses, ballet, elaborate dress-up parties with her dolls and friends. Girls' territory was exciting to overrun from time to time, the rigors of make-believe social rules trashed by gross boys like me. I would delight in ruining tea parties, severing doll heads, and general looting, pillaging, and rampaging. Whatever my sister built was always fun to destroy.

I considered Samantha a perfect child. Her grades were much better than mine, she paid attention in class, and she didn't cause much trouble. But inside she was affected by my parents' divorce. She believed their separation was her fault. She also missed her friends in France. Sometimes she would come home crying. I once asked her what was wrong and she said some of the older kids at Fleming had teased her because she sounded funny. Although her voice sounded normal to me, she spoke with a faint accent. Her friends were similar outcasts, girls who were either too weird looking, too strange sounding, or too socially graceless to be popular. One of her friends lived across the street from us. I could see her apartment from my window, and she was so funny looking—my height, but four years younger—that even I sometimes made fun of her, calling her a hairy tree. I knew that Samantha missed France and the time when she was comfortable and happy, when she fit in at her all-French school where English wasn't even spoken. While I'd experienced those years in France as a rupture that took me away from home, for Samantha France *was* home. French had been her first language.

Although we fought most of the time, Samantha and I shared a secret. She didn't understand what I was up to but sensed I was doing something that my parents, if they knew, would find terrible. From time to time Samantha would come into my room and watch curiously as I got ready for another weekend adventure. She stared quietly as I dug my army jacket out of the closet. It was covered in patches and telltale brown spots from Rush, which like bleach, easily stained fabric. I would give her small clues that I was up to no good. I had to tell someone, and a nine-year-old was a good listener. Underneath my soiled army jacket and my nonchalance I wanted something, someone to offer me an alternative—but this course was hard to leave on my own. My sister seemed to sense this one November afternoon.

I'd bought a package of condoms, or more accurately, the man who sold me pot at the foot of Central Park had bought them for me. I wanted them because my girlfriend, Jennifer, and I had suddenly decided that we should have sex. She also wore an olive-green army jacket and was very thin, with long brown hair and pale skin. Jennifer and I had met at Fleming in fifth grade. Toward the end of sixth grade, as our kissing games became more serious, we decided to become boyfriend and girlfriend. In the beginning it was nothing more than pantomime, a form of role playing where kissing was the treasure we sought.

According to the standard baseball analogy, the farthest Jennifer and I had gone was somewhere between first and second base, an intermediate point where my hands had hovered along Jennifer's bare back under her black T-shirt as we lay on her bed in the tiny maid's room off her family's kitchen. Her lips, red from grape juice, were touching mine as she rested on top of me, her hair tickling my face, and a quiet voice inside me suggested that I move my hands around to the front. I did, slowly, and Jennifer seemed not to mind, but somewhere along the sides of her ribs I stopped, afraid to go further.

My friends and I played fast and loose with the rules. An almost-to-second base became second base, halfway to home. You could, easily enough, round that up to all the way and claim victory: a home run. Our rules bore little resemblance to the original game. There were no referees here, and the spectators received bulletins secondhand, crude facsimiles of the original moves. Home, more than a point, was truly the endgame. Game over. No extra lives needed.

That afternoon, although my hands hadn't the will to go further, Jennifer and I agreed to have sex, a fast sprint from first base to home plate. We would do it the following weekend, but first I had to plan. Two girls from Fleming, where Jennifer still went to school, were planning a party without adult supervision. This would be the perfect place to do it.

According to the books I'd read and health class in school, I knew that Jennifer might become pregnant without birth control. Analyzing the situation, I decided the best thing to do was find an adult to purchase condoms for me. The best person to do that was my drug dealer. Since twelve-year-olds couldn't buy dirty magazines, I was convinced that I couldn't buy condoms either. My dealer didn't dissuade me from this assumption. He was a short, stocky black man with a neat beard who carried his dope in plastic bags hidden between his pants and underwear. He agreed to buy me a box of Trojans after I gave him $10 and told him to keep the change. I could tell he thought the whole thing absurd and an easy way to make a quick $5. We walked along Central Park South past the Plaza Hotel toward Sixth Avenue to a drugstore. He went in. I waited outside, feeling overwhelmed. No turning back. He came out a few minutes later and handed me a brown paper bag, smirking.

At home hours later I showed Samantha the little red box of Trojans and told her that Jennifer and I would have sex later that weekend. Samantha looked at the box, looked at me, and went into my mom's room and told her.

My mom called my dad.

My dad called me.

Seated in the living room hunched over the phone, I listened as my dad told me that I would not be going to the party, that I would not see Jennifer again, and that I would not be having sex now. He then told me to throw out the condoms. We lived on the fifteenth floor and my room faced southward, glass office towers across the way. I went into my room and took the package in my hand. I threw the box out the window and watched as it landed in the middle of the avenue, a little red square on the asphalt. The box disappeared under the taxicabs that, like some yellow contagion, dominated Midtown. Later that night I called Jennifer and told her I couldn't see her again. I would not be going to the party. We would not be going to home base. Home base would elude me for years. I wouldn't reach it until my freshman year in college. I could have told my sister nothing. I could have ignored my dad's demand. When faced with his full attention and rage I did not rebel; instead, I chose to follow. That choice marked my first tentative step out of one adventure and into a new one.

I stopped seeing Jennifer but continued to get high on weekends with the gang. Eric and Tim were trailblazing. Eric claimed a girl he'd met gave him a blow job in a Times Square movie theater. We weren't sure whether to believe him. I decided to one-up the brothers and take LSD. When I told them I would, Eric said, "Okay, we'll watch." Matt agreed. I would take LSD as an experiment, and everyone else would stay with me to see what happened. That Saturday afternoon while we were all hanging out in the park Matt and I asked our dealer for LSD while Eric, Tony, and Tim waited for us on a nearby bench. We found him at the usual spot, near the corner of Fifth Avenue and 61st street, along the path that leads to the Central Park Zoo. He hesitated, then made up his mind, extracting the glassine bags from his pants, fiddling through a potpourri of pills, until he came up with a cream-colored tablet. He charged me three dollars. Matt and I, along with Eric, Tim, and Tony, ran up the hill laughing. I swallowed the pill.

Fifteen minutes passed, another fifteen, forty-five minutes later—nothing! We decided to go to a horror movie. "LSD is lame," I said. "You got ripped off," Matt announced, urging me to demand my money back next weekend. A dozen people were in the theater with us. We were wedged into our seats, feet up on the chairs in the next row, and we did our best to keep quiet while opening bottles of beer. Suddenly the lights were on again and Matt was shaking me. *Wake up,* he was saying, *hey, wake up.* I was so tired I could hardly make my way out of the theater and onto the sidewalk, where the evening light suffused the streets with a muffling peacefulness. Parting ways with the others, I wandered home. I almost fell asleep in my building's elevator. I got into the apartment and went to my room and passed out and slept for fourteen hours. The LSD, I later realized, had been a barbiturate.

The next week John Lennon was shot in front of his apartment house. I cried, and insisted on going to the impromptu candlelight vigil after school the following evening. Snow flurries swirled around the Dakota, dusting its turrets and bay windows, the sidewalks filled with thousands of people. My dad came along, and it turned into a father–son field trip. As we milled around with the crowds at the edge of Central Park I noticed someone I knew—a bearded man, my dope dealer—in the crowd moving toward me. I felt scared, and I grabbed my dad's arm and dragged us deeper into the crowd, trying to escape, moving in and out between people, hoping to make it all seem normal to him. We slipped away, fishtailing into a stream of people holding candles.

For days afterward I played back that moment when my dad and my dope dealer nearly collided. Somehow, without knowing it, in the repetition of memory I was making a choice.

A week later I went into Central Park on my own after school. It was a dreary, cold afternoon and I walked with no destination in mind, drifting along the road toward the Dakota, back to the site of the vigil. In my army jacket pocket I touched a plastic case. Slightly larger than a cigarette pack, hinged, it once held Q-tips. Twenty joints were in there, neatly layered, which I'd rolled myself in the bathroom. The case contained my entire stash, and as I walked down the road that bisects Central Park around 72nd street I thought about everything—about my friends, my dad, myself. I thought about the joints in my pocket.

Up ahead I could turn right into the Ramble, a tightly knotted series of paths through dense woods, following the ridges of Manhattan's short but steep hills. There, hidden by trees, close to the bank of the lake that in summer held boaters and ice-skaters on the coldest days of winter, I could light up, smoke. I passed by the steps that led that way and continued westward. Up ahead I saw a gray metal street lamp with the base panel removed, giving access to the wires inside. I looked around behind me and to either side, and seeing no one, took the case out of my pocket in one quick gesture and slipped it inside the lamp, shoving it deep behind the tangled wires and hurried on. I made a choice— a firm one—to leave the game and find another.

Something inside me grabbed at the clock and turned the hands backward, slowing the dash to adulthood. The choice to retreat was made easier by my dad, who reentered my life after what seemed a prolonged absence. He'd moved into an apartment of his own ten blocks from where I lived with my mom. The move created a feeling of resolution, of a new time coming. The initial shock of no longer being a unified family had ended, and I saw that I'd begun to recover, to find my way between two separate homes. I'd missed my father terribly, and now perhaps I could move in with him for a while. It was he who presented me with a way out. When he began to plan my Bar Mitzvah, setting the date, counting the guests, he offered me a deal. If I would agree not to invite Eric, Tim, Matt, or Tony, he would give me a reward: the gift of my choice. Without hesitation I said a computer.

I fulfilled my part of the bargain by cutting ties with Eric, Tim, Matt, and Tony. When they called I told them that my dad wanted me to hang out with kids from my new school. Far from seeing me as a wimp or a traitor, they understood I was moving on. Tony, the least daring of us, also had begun to pull back, as had Matt. We had gone far and were afraid to go further. That May the old gang wasn't invited to my Bar Mitzvah and I was given what I wanted most—a computer with 48K of RAM, dual floppy drives, and an instruction manual for programming in BASIC. Meanwhile, Eric and Tim joined what everyone in their neighborhood called the 86th Street Gang, a loose alliance of boys who spent their afternoons harassing doormen and pushing smaller kids around. It was minor stuff—shoplifting, chain smoking, graffiti—although some of the kids carried switchblades they'd bought in Chinatown. A year later, though, I heard that Eric was arrested at LaGuardia Airport trying to board a plane to Miami. In his pocket he had several hundred dollars. With him was his girlfriend. He fit the description of a boy who had robbed a deli near Carl Shurz Park a few hours earlier at gunpoint. He was fourteen. Eric went to jail. Tim did too. They won, and grew up first.

4
Extra Life

IN ANTICIPATION OF MY BAR MITZVAH in the spring of 1981 I studied the Torah and the back pages of the mail-order computer catalogs. Both evoked a strange new world of arcane symbols. Each offered the tantalizing possibility of uncovering knowledge buried behind a confusing new lexicon. Aleph, Bet, Gimmel. RAM, ROM, byte. The Torah is the heart of Judaism, our Hebrew teacher told our class of a dozen boys and girls as we slouched in our seats on a Sunday morning wishing we could be anywhere but here. It is the word of God, passed from generation to generation over several thousand years, that gave us our identity and religion. What is holy, above all, are the words, words that could be carried on paper or in our minds.

The Torah is our collective memory, the story of how Judaism came to be; it serves as shared software, booted up through study and reading. The Torah is an algorithm, code with specific instructions for living life. An operating system. So long as we copy the system uncorrupted, verbatim, and pass it forward through time we retain the core of Judaism. Studying the Torah, though, did not give me the same satisfaction as studying *Antic, A.N.A.L.O.G., Softside, Compute!, Joystik*, or *Electronic Games*—magazines I gathered after school and laid about my bed. I read and reread which home computer did what, gleaning from these pages the outlines of hardware, software, programming, and another fantastic symbolic world waiting for exploration. My printed copy of the Torah, a small bound volume, was a chore to retrieve, and my memorization of Hebrew vocabulary words a calvary best done in haste in the back of the school bus at the end of the week. Discerning the relative benefits of the Apple II, Atari 400, Atari 800, TRS–80, and Commodore PET, comparing hi-res monochrome to lo-res color and floppy drive versus tape drive was a pleasure, a joy. At stake here was the ultimate question: Which computer would I choose?

In 1981 there was a swarm of competing home computers, each incompatible with the others. Of these brands only four companies succeeded in attracting substantial numbers of followers; like new religions, each vied for dominance. Each came with its own code, a particular philosophy of computing. For those who lusted after one, the machine's style and substance were embraced as a reflection of ourselves. Just as kids two generations earlier had lusted after particular cars, each with its own internal logic of cool that said "this is who I am and what I stand for," and the generation after that had followed different musicians and sounds, so we followed different computer makers.

The four companies with substance and verve were Apple, Atari, Commodore, and Radio Shack. Boys who came to computers through electronics wanted an Apple. Boys who came to computers through video games wanted an Atari. Boys whose parents wanted them to have a "serious" computer got a Radio Shack TRS–80. The Commodore PET was a graphics machine too, but I didn't know anyone who bought one. In the far distance was a new computer, the IBM-PC, released in April 1981. That machine was not for kids. It was for grown-ups, and none of us ever dreamed of owning one. There was a purity to the market driven foremost by the exuberant joy of hobbyists and children. It was untamed, undisciplined by serious uses such as accounting and word processing. These machines weren't for work, they were for play, for exploration, for adventure. What they were for was not up to marketing experts and advertising agencies to decide; it was up to us—the millions at home who took to programming, turning what once had been the arcane art of scientists and graduate students into a nationwide pastime.

By the end of seventh grade I'd made several new friends. All were new kids like me, and none did drugs or had much idea of what I'd done over the past fall. Victor, Aaron, Scott, and Kenny instead offered me the comfort of being with kids whose sense of daring was winning at strategy and video games, or discussing the merits of different war movies, guns, and explosives. Most important, several of my new friends owned home computers. Kenny, Aaron, and Scott acquired machines in seventh grade. The TRS–80 for Kenny; the Apple II for Aaron, and an Atari 800 for Scott. As seventh grade waned I studied each machine judiciously, comparison shopping, feeling the energy, searching for the one that would match my own imagination.

The dream that lured me to computers was the fantasy of parallel universes, of an escape into a reality that could be animated and made real by a computer. I wanted one that could draw beautiful worlds, create vivid lands for me to explore. The ur-fantasy my friends and I shared came from science fiction: the scene in *Star Wars* when Luke Skywalker and Chewbacca play chess as their spaceship, the *Millennium Falcon*, races toward the hidden rebel base. The chessboard is circular and the pieces are miniature three-dimensional holograms of monsters, walking, alive. Luke moves a piece to take Chewbacca's and it hops over an ugly orc or troll, landing in an enemy square and killing its opponent with one hard blow, blood lust on its face. That moment evoked the rich, fantastic illusion we so desperately dreamed our computers could craft. No such machine existed then (or now); but the desire to weave realistic illusions remained, sated by what we could have: 8-bit graphics, 255-bit monsters built of tiny two-dimensional rectangles, pixels on a screen.

*   *   *

I was a graphics junkie. What I cared about most were pictures, drawn fast and beautiful. I wanted a machine with color, sound, and speed. I wanted a machine whose graphics mesmerized. Graphics, like a book or film, created the illusion of another world. Well-done graphics, like a well-done book or movie, suspended disbelief and brought us out of our bodies into a new place. This was why I sought out my first computer. Which could make the best worlds? Which would take me there fastest?

Kenny, short with neatly parted black hair and big brown glasses, lived (as most of us did) with one parent, his mother. His apartment was twenty blocks north of mine, near Aaron's, and there Kenny had his TRS–80—the Trash Eighty, as we called it. Kenny's mother bought it for him. The computer came with 16K of RAM, a tape drive that used audiocassettes to store programs as audible beeps and bleeps, and a built-in monochrome monitor. Kenny's Trash Eighty sat on his desk catty-corner to his bunk bed where sometimes I would sleep over. The Trash Eighty had its charms—cheaper than the rest, rugged, and easy to use. Built by Radio Shack, a division of Tandy, the TRS–80 was presented as the machine for adults who wanted a home computer for business or personal finance. A low-end TRS–80 could be had for $499.

Huddled together in front of Kenny's desk, we played games he'd bought for the Trash Eighty. The machine had no color and the sound was lame; Kenny's games weren't impressive. Worse, his mother wouldn't buy him more games. She wanted him to use the computer for serious things like typing, math, and spelling. One afternoon when we were taking turns at a game Kenny's mom came into the room and said to Kenny, "How much longer are you going to sit in front of that thing? Why don't you go play outside?"

"Ma," Kenny said, "I'm busy."

"Oh!" his mom said, "too busy even to listen to your mother?"

Kenny kept playing.

"Look at me when I'm talking to you, young man! Too busy for your own mother?"

When Kenny's mom got mad her voice became very high, almost shouting. I started to squirm.

"I want you two to go outside and play. Now! Do you hear me?"

"Ma. LET ME FINISH MY GAME FIRST!"

Kenny's mom started to twitch like she was about to smack him. *Uh-oh,* I thought, starting to roll out of the way on my chair as his mom strode forward toward Kenny, and before I could duck she'd reached over and unplugged the computer.

"MAAAAA!"

"That's it," she said. "No more computer! You're grounded for a week."

"But Ma—"

"You want two weeks? You want me to call your father?"

That was it—the ultimate threat for most of us with separated parents: mom calling dad. Kenny slid off his chair, defeated. We went outside, walked to Radio Shack, and looked at more TRS–80 gear in the store, stuff that Kenny was even less likely to get now—rack after rack of audiocassettes in glassine packages, each containing an audio recording of bits and bytes, programs stored on acoustic tape. Kenny had an anemic collection of microcassettes neatly stacked in his desk drawer. Loading and saving games took a long time. The best games Kenny had were Space Invaders and Breakout, a Ponglike game whose object was bouncing a ball off a wall until all the bricks are destroyed and a new wall appears. Both games were designed by kids a few years older than us in Great Neck, Long Island, whose company was called Software Innovations. Their oldest executive was fifteen. This computer craze—unlike earlier crazes such as late-night radio, LP rock albums, or drag racing—could pay off with rich dividends. Earlier crazes were run by adults: record labels, film studios, General Motors, intrepid travelers bringing pot up from Mexico. Growing up with computers when computers were still young was different.

Here you could do more than just consume what adults passed down to you, cash registers *ka-chinging* in ecstasy. For every pocketful of quarters dumped into an arcade game and preadolescent savings bond cashed in for a home computer, somewhere a kid programmer was writing code, making software, and passing it on to friends at school, or—if the product was good enough—getting it brokered through one of the grass-roots cooperatives formed to distribute software, later known as "shareware." An enterprising kid could write a program, give it to a co-op who would distribute it for free, and if so moved (as a kind of honor system) ask for $10 or $15 to be sent by mail. Two contradictory interpretations were made by the media, which looked at computer programming and video game addiction with puzzlement. The first was one of initial protest—kids were rotting their brains on mental junk food and becoming video game truants; arcades were swamps of iniquity, breeding grounds for vandalism and nihilism. The second was a tale of superbrains, *uber*kids with great round glasses, faces basking in the late-night glow of a monitor, coding. Both tropes were linked to the same root: these kids were amoral addicts. Videojunkie just wants a fix and cares little about anything else—except perhaps pizza. Superbrain's addiction is the thrill of power. Where Videojunkie thrives on a high score, Superbrain thrills at the idea of hacking into NORAD. He's unstoppable because grown-ups can't think like he does. We are at the mercy of his mood. Maybe he'll bring down the national telephone network today . . . or not. Such images bore little resemblance to what many of us experienced—the thrill of pioneering.

As computers entered our homes we were defining a new culture through gleeful experimentation, one that with the Internet in the 1990s would become dominant, capturing as much attention as did rebellion in the 1960s or jazz in the 1920s. Yet where the latter movements began from the top with supremely talented individuals and trickled down, digital culture began at the bottom and trickled up, starting in cramped bedrooms like Kenny's and moving upward from kid to kid until it colonized the outside world. Propelling its movement were two factors: the availability of new technology and our natural desire to grow up into men. For boys going through the transition from teenager to adult, we wanted what previous generations wanted—to be different from our parents, to have separate identities. In the computer we found a devoted accomplice. We could help define it while it helped define us. For a generation in which everything seemed to have been done before, what was there left to do? Drugs were done, music was done, street revolution was done. Everything seemed old. Except this. Together, computer and kid co-existed in a golden age, a time when the machine was available to us unconcealed, stripped to its component parts, when adults barely understood what we were doing and the outside world did little to interfere with our probings and pokings. For the Atari generation the evolution of the machine briefly matched that of our adolescent selves, becoming a vessel and partner, a coconspirator in our mutual coming of age.

Aaron lived with both of his parents, a few blocks away from Kenny. Aaron's dad did an arcane thing called "finance," which none of us understood. His dad would sit in a little room and use a computer to make charts. He made money from the patterns that emerged. Later Aaron explained that his dad was an arbitrageur, which meant he bought things low and sold them high. Aaron's dad had no fear of technology. He was the first to buy any new high-end electronic device—calculators, digital watches—the kind of executive from whose office in 1972 I might have stolen something digital at a cost of several hundred dollars. So it seemed perfectly natural for Aaron's dad to buy an Apple II computer.

The Apple II was a computer for cool grown-ups, science whizzes, and kids who liked to take things apart and rebuild them. From the beginning Apple's founders, Steve Jobs and Steve Wozniak, had insisted that all technical details of how the Apple worked—such as how the hardware communicated with the software and what codes were controlling the chips inside—were available to anyone who asked. The design of the Apple II reflected their ethic: it had an easily re-movable top panel so that you could fiddle around with the circuits inside. The Apple was built on the virtue of Information Transparency—the more people knew how they worked, the more programs would be written; the more programs were written, the more people would want an Apple. A virtuous cycle. And it worked—the Apple had more software from a wide range of authors than any other home computer.

The problem with Aaron's Apple was with its games; they lacked redolent color, verisimilitude. Apple games were crippled because the Apple II's microprocessor, the Motorola 6502, had no support. All graphics and sound had to go through its tiny logic gates. That made sense for a machine whose designers saw it as a home computer, not a souped-up game console. Their ideal user was made in their image—a fearless, curious hobbyist willing to get down with the machine, maybe write some code of one's own along the way. I wanted raw power. Noise. Explosions. Action. I wanted to save the world and then hit RESET to give it another try. Scott could do that. He had an Atari.

Atari. Even the name was cool. It meant "check" in Japanese, as in chess, but we didn't know that. It just sounded good. Scream *Atari* as loud as you can, do it again, and you'll know what I mean. And the logo—three vertical parallel lines with the outside lines curved slightly outward at the bottom—created an image that said: Alien. Alien attack. Like in Space Invaders. Atari made games, and games were the best thing ever. Apple? A fruit. I liked candy, ice cream, and anything made by Sara Lee. Fruit and vegetables were hated, parentally mandated food groups. A computer for kids should not be named after such beastly things. So Apple had a few handicaps to overcome before an arcade fanatic like me would consider owning one.

Scott was an arcade fanatic, a console cowboy, a boy who lusted for better, faster graphics. Whenever we had free periods between classes Scott and I would head "down the hill" to play video games. Scott had the frenetic power and charm of a kid on a constant high-energy, hyperactive quest. His bookbag was always stuffed with the latest computer magazines; his notebooks were covered in elaborate doodles—warriors, guitars, battleships, sprawling scrawls of shapes. A head shorter than me, Scott also wore glasses like the rest of us—brown plastic frames, the lenses cloudy and scratched, never cleaned since the day they left the optician. Going down the hill was a release and a stimulant all at once. "Up the hill" was Horace Mann, overlooking Van Cortlandt Park in Riverdale. Unlike Fleming, Horace Mann is enormous, with plenty of greenery; it resembles a suburban high school, with a football field in the center bounded by faux-gothic and Georgian-style buildings. At the far end of the campus are tennis courts, and the gym building contains an Olympic-size pool. Up the hill is cloistered, private. Down the hill is chaotic, public. This was the province of kids who loved arcade games, under the shadow of the elevated Number 1 IRT subway terminus, its metal staircase alighting a few steps from Joe's Pizza. Joe's was a dozen or so yards from the Deli, and both contested for our attention. Joe's offered Tempest, Battle Zone, and Sea Wolf. The Deli had Donkey Kong, Defender, and Missile Command. While the rest of the school sat in class, in the library, or gossiped in the cafeteria, we battled.

I would play my favorite arcade game—Tempest—at Joe's. Built by Atari, Tempest was one of the first video games to offer a real sense of three dimensions and fluid speed. I took on the role of a claw-shaped object on the rim of a wire-frame tube, trapezoid, or combination of both.

Using a spinning knob to whirl around the rim, I hurled bolts of fragmented energy down at crawling objects heading up the wall toward my perch. If they made it to the top I would be crushed, eaten by one of these odd bugs. The finest moment in the game came after clearing an entire level of parasites. At that moment I hurtled forward, down down the electric tube and out the other end into space toward an approaching object—the next level—on whose edge I would land and begin again.

I mastered Tempest, cycling onward to higher and higher levels signified by changing colors—blue, red, green, yellow, crystal white. Easily earning extra life, I became so close to the machine that thought itself subsided; everything was instinctual. A slice of pizza perched on top of the machine, the spicy scent of pepperoni mixing with sips of icy cold, sugary sweet grape drink. Lines of quarters preserving my lease on the next game. A cluster of people leaning over my shoulders. On our way up the hill, moments before history class, Scott and I would quickly run through our latest conquests, moments well played, new levels breached. I would think about Tempest, replaying moves, seeing a ghost image on the back side of my closed eyes, relishing the exquisite graphics. Minutes later I'd be listening to Mr. Newcombe, my history teacher, spinning long tales concerning Herodotus, the Peloponnesian War, and the *Iliad*. I rarely took notes. Instead I'd be scribbling pictures, sometimes from Tempest, or a hybrid in between Mr. Newcombe and the game—Peloponnesian warriors perched on the edge of an electric-blue abyss, hurtling lightning downward into the pit.

*   *   *

Games. Scott's room was littered with them—cigarette pack–sized plastic cartridges, each containing a circuit etched with the software of a tank battle or alien swarm waiting to be loaded. Floppies slipped across the yellow shag rug, wedged behind the desk or somewhere beneath a pile of back issues from *A.N.A.L.O.G.* and *Joystik!* Manuals cascaded out of overstuffed desk drawers, pages poised to fall, jammed between the edge of the closed drawer and the desk frame. In the place of honor was Scott's Atari 800 connected to a color television set, a disk drive, a cassette drive, and four joysticks. His small room was off the kitchen, a fragment of the living room intended as a dining area, now bordered by the kitchen on one end, a window opposite, and a plastic sliding partition that divided Scott's room from the living room on the other side. Inches away, separated from us by that makeshift wall, Scott's dad would sit on the couch watching marathon sessions of the Knicks. The screams of the crowd and the announcer intoning the progress of the latest basketball game came clearly through the plastic between us.

Scott's dad was a hardware hacker, the man behind the Big Muff Pi, Screaming Tree, and Hog's Foot—sound distorters celebrated by rock 'n' roll musicians worldwide. Plug the Big Muff Pi into an electric guitar and the sound waves contort, knot together, and emerge as a way-new sound of thrashing, classic rock from the seventies. Scott's dad built and sold this stuff through his company, Electro-Harmonix, housed in a warehouse on 23rd street. Scott was something of a hardware hacker too. His walls were lined with plastic circuit boards, glued there as decoration. Some had keyboards or keys from cash registers. In Scott's room, crowded among magazines, floppies, cartridges, and circuit boards we would sit hunched over the TV and play and play, while Scott's dad channel-surfed between matches and we waited between games to raid the kitchen for the next round of Twinkies, Hi-Hos, and cupcakes.

Linked to our brains via the black plastic cables, joysticks in our hands, and the sugar rush were four chips—the 6502, ANTIC, POKEY, and the GTIA. Atari had extended the capabilities of the 6502, the same chip that powered the Apple and the Commodore PET (TRS–80 used a Zilog Z–80 as its brain), by adding three additional chips—ANTIC, POKEY, and the GTIA. Where Apple's computers struggled to refresh the screen and display blazing color, these three chips served one combined purpose: to make the Atari 800 more responsive by taking pressure off the primary 6502 chip. Like dedicated servants, ANTIC, POKEY, and GTIA specialized in graphics, freeing the 6502 to do what it did best: computation.

When Scott and I booted up a game this confederation of chips acted as a single unit, speedily passing data from our fingers—the motion we chose coming through the joystick cables—into the computer. From there POKEY grabbed the information. POKEY was a specialized chip that handled input-output, sensing when a button or key was pressed. POKEY passed that data to ANTIC fast, and the 6502, the center of all operations, could continue overall computation without breaking stride to process the new input. ANTIC translated the press of a button into the appropriate image—a missile launch perhaps, or a jump across a pit. The acronym stands for Alpha-Numeric Television Interface Circuit, and sixty times a second ANTIC turned all the inputs into a stream of graphic commands, updating the game world, ready for handoff to GTIA. GTIA took ANTIC's graphic commands and translated them into television-ready signals. This exquisitely rapid, coordinated ballet took place much faster than an eye blink and produced a visual environment whose responsiveness surpassed anything Apple, Radio Shack, or Commodore offered. Aaron and

Kenny's computers couldn't compete with that. Atari was the computer I wanted.

In May I walked down the center aisle of the synagogue holding the Torah, wrapped in a linen shroud, and decorated with stars of David woven in gold thread. Together at the altar the rabbi and I unfurled the Torah to the right passage, which I read out loud, using a silver pointer to trace the rhythm and flow of the Hebrew words. Aaron and Kenny were seated in the audience in rows of pews, along with my other new friends. Later that day we gorged ourselves on chocolate cake at my aunt's house and watched the movie *M*A*S*H* on television using my dad's VCR, which he'd brought over special for the occasion.

My dad knew that I wanted the Atari 800 to play games. I consumed pocketloads of change every week. But he felt that was all right. Like me, my dad was into gadgets. He had a VCR in 1978, a monstrous steel box that weighed seventy pounds. He'd had cable TV in the sixties, when Manhattan was one of the first places in America to get wired. So when I asked for a computer, part of him instinctively loved the idea. A computer at home! He also intuitively grasped that computer games could serve as a gateway leading to something else: a productive fascination with mastery. My dad was smart that way. He wanted me to be good at something positive. He had no idea what form that would take or that I'd develop a code of my own, an unspoken set of Commandments brought out by exposure to computers.

Computers do think a certain way, and exposure to the way computers think in turn changed the way we thought. They programmed us as much as we programmed them, and thus a generation came of age experiencing not simply a new toy but a different way of seeing. These unspoken, unwritten values grew spontaneously. No one directed them. No one told Aaron, Scott, or me what they were. We just felt them. The First Commandment was that computers make the world a better place. Second, kids should have access to computers, and any grown-up who got in our way should be mistrusted. Third, kids were duty-bound to share computer information with other kids—knowledge had to be passed on; computer secrets were a cardinal sin. Fourth, programs were owned by everyone and should be shared and improved by all. Fifth, all that mattered was your computer knowledge, not what you looked like or where you came from. Sixth, all exploration is good. If you believed in these Commandments you were part of a new tribe, what some would call hackers, others simply computer kids.

I waited four long months before the Atari came. My dad wanted to get the machine wholesale. For gadget lovers, part of the game is getting the gadget cheaper than anyone else. Paying full price is unacceptable. Thus the delay, as he worked the system. I survived. Summer intervened, and with it a return to camp, to Vermont, where away from video games my computer pangs subsided, ameliorated by the trunkload of books I brought with me and by the games we played outside in the sun. When I came home, poised to enter eighth grade, I moved in with my dad in his new apartment, where he lived with his girlfriend, who two years later became my stepmother. My sister stayed with my mom. There, in my new room, my dad and I discussed drilling holes in my bookcase so that power cords could snake their way through and connect to the computer, which would lodge in a specially built area in the center of the bookshelf, waist-high so I could sit in front of it and play.

When the Atari 800 finally came in late September by mail, my dad and I tore open the package. Inside was a big white box. It had a plastic carrying handle on top and a picture of the Atari 800 on its side. Inside, wedged between Styrofoam pads, was the caramel-brown case of the Atari 800, with darker chocolate-brown keys and a row of orange and yellow buttons along its right side. My machine came with 48K and a few game cartridges that fit inside a specially designed panel that flipped open on the top of the computer. We snaked the cables through the shelves, lay the computer in its special spot, attached it to my color television, flipped the On switch in the back, and tuned the set to Channel 3.

The screen was blue, a rich bright blue. A phrase ran along the top in big, white uppercase letters easy to read on a screen:

`ATARI MEMO PAD`

I pressed the keys and letters appeared on my television set. The same letters I was typing. The Atari made a little synthesized clicking sound each time I hit a key. The Memo Pad program was etched in the Atari's ROM, firmware, proof that the machine was working. Running a game meant inserting a cartridge in the top slot and restarting the machine. Each cartridge contained a ROM of its own, and once in place the Atari would boot the program on the guest ROM, pushing the Memo Pad aside in favor of Pac Man or Combat, or whatever game cartridge was on hand. In back was a special port, a hole designed to accept a cable that connected to a disk drive. With a disk drive, an optional add-on, the Atari became more than a game machine—it became a programmable computer. It opened a world of games, games that came on floppy disk. Games that could be saved and altered if you knew the codes. So I was dou-bly delighted when my dad showed me the disk drive he'd bought to go with the Atari: the one I wanted. The one I'd researched in the back pages of the computer magazines. The Percom. A heavy thing, far heavier than the computer, the Percom accepted two 5.25-inch floppy diskettes in parallel slots—I could read from one and write to the other. Each disk stored 90K on a side for a total of 180K. We plugged the Percom in and booted up the disk marked ATARI DOS II.

An operating system.

There on my desk sat a full computer. A machine that five years earlier hadn't existed. The Atari came with another cartridge labeled PROGRAMMING IN BASIC.

A programming environment.

I didn't know how to program, but I wanted to. In the magazines I read many of the games came as BASIC programs, typed directly as code onto pages and pages. Scott told me how he'd sit and type these programs into the Atari in BASIC, save them on the disk, and then have a game. He could even change the way the game looked and played by altering the code. He could give himself more missiles, extra lives, or make the aliens come down differently, changing their appearance. I wanted to do those things too; I wanted to see how games worked. Holding my Atari instruction manual in one hand and my BASIC cartridge in the other, I was ready to begin.

5

Atari

THE FIRST NIGHT with my Atari I sat on my bed surrounded by books and magazines. I had a lot of work to do. Outside, the sun had set. Inside, my new computer shone, illuminated by the lamp next to my desk. I kept looking up, just to make sure it was still there. The sharp smell of brand-new plastic had faded now to a trace odor that filled me with anticipation. A few copies of *SoftSide* magazine and *A.N.A.L.O.G.* were scattered around me. I focused on what I'd discovered to be the key text—the manual for Disk Operating System II.

Published by Atari, Disk Operating System II described how the different parts of my computer system worked together and how I could control them through software known as ATARI DOS II, or DOS for short. I wished at first for one or two games to play, like Centipede or Space Invaders. Instead, I found a new kind of game in two programs—ATARI BASIC and ATARI DOS—that came in ROM cartridges. Here was a different game, a good one: discovering the secrets in my new computer.

The Atari manual explained how to connect the disk drive to the computer and use the operating system to run files off floppy disks. Within a short while I was loading and copying files. The Atari gave many such payoffs for my ef-

forts. I connected with the machine quickly; the barrier between me and its circuits was low, low enough for a thirteen-year-old to start programming right away, even if I had
little idea of what that word meant. That night I wrote my
first program pretty much by accident, for fun—and to
show off for my dad.

READY

The word, in stark capital letters, stared patiently at me
awaiting my command. READY was a word I would come
to know well, the universal symbol for hello between computers and kids during the heyday of the home PC. READY
meant what it said: I am ready for you, do what you want,
ask what you will. I inserted the BASIC cartridge and its
ROM pushed the default MEMO PAD out of memory when I
restarted the computer. I didn't understand it yet, but
READY meant that I was inside a programming environment. All 48 kilobytes of my Atari were focused on me.
What did I want it to do?

Simply by playing games I'd learned a few BASIC commands such as LOAD and RUN. LOAD retrieved the code
from the disk and RUN executed it, turning its lines into instructions that the Atari's 6502 chip could process. Since I
had nothing to LOAD, I would need to write my own BASIC
program. The manual described the PRINT command.
PRINT in BASIC made things appear on the screen. For example: PRINT "HELLO."

I typed that in, below READY. PRINT "HELLO" and then I
pressed Enter.

"HELLO"

READY

The gamester in me moved slowly, pacing myself. Early on I'd played games impetuously, wanting to master them right away. Now in eighth grade I was learning the fun of teasing out the game, enjoying each level before storming on to the next.

PRINT "HELLO THERE"

```
HELLO THERE
READY
```

My first program. PRINT "HELLO" became the first line of code—the first of thousands, perhaps tens of thousands—that I would come to write over the years. I moved to the next level.

Such one-line commands treated my Atari like a machine without memory: the moment I pressed Enter, PRINT "HELLO" was executed, then forgotten forever. The strength of my computer, like all computers, was its ability to store information. Storing more than a single line of code required a different format. You had to begin the command with a "line number," like this: 10 PRINT "HELLO". After hitting Enter the computer would not display

```
HELLO
READY
```

Instead, it would patiently wait for another line, such as 20 GOTO 10—which is what I typed in. My Atari waited, ready. I typed RUN and pressed Enter.

```
HELLO
HELLO
HELLO
```

The words flowed down the screen. It would go on for-
ever. This was my first *stored* program. Two short lines.
Powerful. Forever in two. The GOTO 10 command told my
Atari to jump back to line 10 and PRINT "HELLO" again,
jump back to 10 and print, and again and again. HELLO
HELLO HELLO. I hit the key with the word BREAK on it.

STOPPED AT LINE 20
READY

Patient again. Waiting. I was ready. Eagerly I jumped
ahead, typing this in:

```
10 DIM NAME$(255)
20 PRINT "HELLO THERE. WHAT IS YOUR NAME";
30 INPUT NAME$
40 PRINT "NICE TO MEET YOU, "; NAME$
50 GOTO 40
```

It took a while to get it right. I was using something called
strings, variables that stood for words. The $ meant a vari-
able was a string. It made sense: $ looks like S. String. I
liked the word. DIM made NAME$ a little spot in memory
255 bytes big. Each byte was enough room for one letter.
Each letter had eight bits in it. Each bit made a pixel shine
on my screen. Bit-byte-kilobyte. Easy. The books told me.
255 letters. That's a big word. 1,024 letters makes 1 kilo-
byte. 8, 255, 1024. Patterns. 10 DIM NAME$(255) NAME$
was your name. WHAT'S YOUR NAME?

David
NICE TO MEET YOU, DAVID

HAL, see what I have here? I have my computer, and it does what I want it to.

f u c k
NICE TO MEET YOU, FUCK

Anything I wanted.

"Dad," I yelled, "come in here." I hit Enter lots of times, until FUCK disappeared, replaced by READY.

"Come in here!" I yelled again. Where was he? I got up and went into his room. He was lying on his bed, reading *Car & Driver*. He's extremely nearsighted, and the magazine almost touched his nose. His girlfriend, Joanna, was next to him, reading a book. "Come look at this," I begged.

"What?" My dad looked at me, annoyed.

"I did something on the computer."

This got his attention. Earlier, when I first set up the computer I nearly had to push him out of the room, he had been so curious.

"Oh good." He turned to Joanna, touching her hand, and said, "Let's go look."

"You didn't bring down the Pentagon or something while we were lying in here, did you?" she asked, grinning.

"Not yet," I replied. "Later."

We came into my room. "Type RUN," I told my dad. He leaned over my desk with his face close to the keyboard and typed. My computer responded.

HELLO THERE. WHAT IS YOUR NAME?

"Type your name," I said. My dad grinned.

NICE TO MEET YOU, MICHAEL

```
NICE TO MEET YOU, MICHAEL
NICE TO MEET YOU, MICHAEL
```

It went on and on.

"You did this?"

I nodded. "Yeah. I read the book," I said, pointing to the manual. We stared at the screen as the text flew by in an endless loop. They were mesmerized. I hit BREAK suddenly.

```
STOPPED AT LINE 40
READY
```

"Let me try," Joanna implored. She leaned over and typed RUN without hesitation. Then she typed in her name.

```
NICE TO MEET YOU, JOANNA
```

"Hah!" she said, watching the phrase scroll across the screen. She made this little laugh—"hi-hi-hi"—and then "that's great."

She turned to my dad. "Did you see what your son did?"

"That's my son," he said, beaming.

I was starting to feel pretty chuffed myself. I preened a bit, typing LIST so they could see the program's code.

"See these are line numbers and this takes your name and this prints it out and then goes back up here and prints it again and again until you stop it," I explained. My dad and Joanna followed the lines. "If I add a semicolon and an empty space after the part that prints out your name, then the words come out on a long line." I edited the line quickly, turning it into

```
PRINT "NICE TO MEET YOU, "; NAME$; " ";
```

"Now try it."

My dad typed in his name again.

```
NICE TO MEET YOU, MICHAEL NICE TO MEET
YOU, MICHAEL NICE TO MEET YOU, MICHAEL
NICE TO MEET YOU, MICHAEL NICE TO MEET
YOU, MICHAEL NICE TO MEET YOU, MICHAEL
NICE TO MEET YOU, MICHAEL NICE TO MEET
YOU, MICHAEL NICE TO MEET YOU, MICHAEL
NICE TO MEET YOU, MICHAEL
```

The words formed a pulsating mosaic of white letters suddenly abstract, hypnotic. As the letters rolled by and silent praise filled the room everything around me took on a peaceful, focused feeling. I started to tingle. "This is the best present ever, dad," I said. I smiled and he put his hand on my shoulder. That was as much emotion as I wanted to share with him. I was in the midst of a long-running punishment loop: prohibiting my parents from getting to know me in retaliation for having my sister (I was no longer the center of attention), taking me to France, and getting divorced. Being open with them would have been too generous. They had to suffer first, and the best way to accomplish this was not to let them share in the joys of my life.

I could tell my dad wanted to sit down at the machine, but this was my territory. I was better than them at something—better than adults. I didn't understand that was why I was so happy—that they couldn't do this and I could. I loved my computer. I wanted to do more, and not realizing that part of my pleasure came from sharing this with my dad, I sent him and Joanna out of the room.

The world wanted us to do more, and gave it to us. Millions of 16K, 32K, 48K, and 64K computers came into our homes. 1981, 1982, and 1983 would produce an extraordinarily rich sandbox built of bits and bytes for the collective will of thousands of boys. This was a vernacular revolution—kids teaching kids. Learning by doing. Roger, Scott, Aaron, and I traded information. One summer between eighth and ninth grade Scott stayed home and programmed every day. He wrote a program for his dad's new product, the Alien Group's VoiceBox, a speech synthesizer that his father assembled in his warehouse on 23rd Street. For Scott's dad—a perennial entrepreneur with his eye on the next big thing—business in the '70s had come from rock 'n' roll: electric guitars and the Big Muff Pi. In the '80s he saw it was computers: software and the VoiceBox. The size of a deck of cards, the VoiceBox was a black box with a cable coming out one end. Plug the box into the Atari or an Apple and your computer acquired the ability to speak. Type in words and the VoiceBox would read them back aloud. Scott wrote the software to control the box, to interface between the 6502 and this new device. He drew a man's face on screen that mouthed the words. I got one for free in ninth grade.

It made perfect sense for Scott to program the VoiceBox. Who else could? Much as each generation thinks that they alone discovered sex, we thought that computers were ours, that they'd arrived from outer space and landed at our doorsteps just for us to play with. Few of us knew where they came from or that we'd hijacked an attitude from hackers, hobbyists, and hippies who discovered computers in the late sixties and early seventies, and that in turn these computers had been created by iconoclastic, freakish engineers and grad students before them. Few of us believed that we could learn programming from grown-ups or that some of them could know more about computers than we did. There weren't too many grown-ups around like that.

Where would we meet them? For most it seemed we were
on our own. Those who met these older masters would discover something more about computers, something deeper,
rooted in decades of endeavor, something I would come to
appreciate much later in life—an understanding of how
these machines were passed from one generation to the
next until, having mutated along the way through luck and
coincidence, the computer flowered and matured at the
same time we did.

Our lack of understanding about the history of computers
reflected in part the sense of ennui children feel when they
learn about ancestors. Adults weren't much brighter about
it either. Popular culture was keen to celebrate the arrival of
the computer as a bolt from the blue, a mysterious, unnat-
ural force, intrinsically malevolent. The few films that de-
picted computers bore little resemblance to what we
experienced, although they represented alternatives that, if
possible, would have been calamitous. *Tron*, a Disney
movie released in summer 1982, told the story of a man
trapped inside a computer, fighting for his life to get out. It
offered 53 minutes of hi-res graphics, a breakthrough in 3-D
modeling and an early example of how computer animation
was rapidly evolving from crude wire frames to lushly textured objects. My friends and I, however, found the film
painfully stupid and clueless; it lacked "street cred." The
jargon was all wrong, the computers ran off implausible operating systems, and the protagonist was a thirty-something
engineer who was supposedly a whiz at computer games. A
better creation from the adult world was *War Games*, which
came out in 1983 and featured a kid our age, fourteen or so,
who breaks into a top-secret military computer using his
home computer and modem. With his clunky IMSAI computer running the CP/M operating system, five-and-a-quar-

ter-inch floppy disks, and modem with acoustic coupler, the hero in *War Games* had street cred.

While the kid is in school, his computer "walks" through random telephone numbers, dialing each one, looking for another computer at the end of the line. When it hits, the number is stored in a special file. At night the kid calls up the hits to see what's out there. One evening he stumbles on an intriguing machine that he thinks belongs to a game company. Upon connecting he sees the following words on screen:

SHALL WE PLAY A GAME?

A game? Yeah. I think so. Which one? The choices scroll down. How about GLOBAL THERMONUCLEAR WAR? He picks that one. Unfortunately, it's a secret military computer and it's not playing a game. The kid unknowingly sets off a countdown to the real thing, rockets fueling up somewhere in the Great Plains, Strategic Air Command radar screens going haywire. Seated in the theater I could feel the insane plausibility of a computer with a mind of its own while the world teetered on the brink of Mutual Assured Destruction. In the movie's end sequence the boy stands in the control room at NORAD watching the adults struggle vainly to switch off the computer's launch cycle. Meanwhile, nonexistent Soviet ICBMs are drawn as incoming lines on the overhead map of North America in the grown-up version of the game Missile Command. Soon, however, these lines will pass the point of triggering an automatic U.S. retaliatory strike—a strike that will be all too real. How can they tell the computer to stop? Sensing the answer, the kid suddenly hits on an idea: run a different game—not GLOBAL THERMONUCLEAR WAR, but TIC TAC TOE. No one but the kid-like professor, who wrote the computer's program, understands the brilliance of this. As the computer plays itself at Tic Tac Toe, iterating thousands of games in seconds and filling the overhead screens with a dazzling array of Os and Xs, it quickly becomes apparent that there is no way to "win" and that two experienced players will always play to a draw. Likewise, there is no way to win at Global Thermonuclear War. On its own, the computer chooses to cancel the launch command. The message of the movie is clear: only the kid can stop this, because only a kid can truly understand the mind of a computer.

War Games spoke to our desire to be heroes and amplified it to the extreme: saving the world from nuclear holocaust. The adults who wrote the movie clearly understood the Six Commandments. The message is one of hope. The military represents an adult world based on secrets and hierarchy, whose hidden command post and foolish generals stand in contrast with the world of the kid. For him the whole idea of secrets is antithetical to his way of life, symbolized by the free flow of information coming through the modem in his bedroom. Even though the kid almost destroys the world, his intentions are innocent—to learn, explore, play. The sentiment is clear: Were the military run by kids like that, there would be no threat of war.

War Games is fiction, but in real life breaking into computers was tantalizingly simple. A year before *War Games* came out my dad unknowingly helped me break into my first computer. Within weeks of starting to use my Atari I'd collected dozens of games, programming books, stuff swapped in school, in the cafeteria with other kids who had home computers. My dad had become alert to my computer needs; he gained vicarious pleasure from my satisfaction. One evening at dinner he asked, "Do you want a modem?"

I looked up at him. I couldn't believe it. Every kid loves to get presents, especially big ones when they're not expected. How fast? 1,200 baud, he told me. He knew the speeds. 1,200 baud was fast, really fast. Each baud is a bit: 1,200 baud means 1,200 bits go by in a second; 1,200 bits means 120 characters in one second; 120 characters is a lot of code. With a modem my dozen games would be nothing, a puny subset of a huge collection. My new friend Roger had a 300-baud modem. No one I knew had 1,200 baud. My dad said there was an extra one at his office. Free. I could have it if I wanted it.

The next day after school I went to my dad's office in Midtown. The modem was stuck on a shelf in between boxes of paper over a small photocopier. My dad started to explain why they didn't need it but I wasn't listening. He pulled the modem down and handed it to me. Heavy, shiny metal with a red panel in front and diodes behind, a plug, and a manual; it was made by Hayes. Mine—1,200 baud. But will it work with my Atari? I wondered. My dad told me to figure it out.

I'd gotten good at figuring these things out. Patterns of two. Once you cracked the system you could figure out how printer speeds, modems, disk drives, and even the speed of the computer were related. The whole thing is one big system, I was starting to see. That's what digital is all about: being related. These were not separate technologies in the way analog technologies are. Instead these technologies formed a class of objects all based on the same binary duality of on–off, 1 or 0. Experience on a handheld game, arcade console, and home PC added to an overall understanding of all things digital. In the school cafeteria I talked to Scott and Roger about my new modem and Roger told me I needed a special box called the Atari 810 to hook it up to the computer. The 810 also allowed a printer hookup, as well as extra disk drives. It was an I/O box—Input/Output. Shaped like an oversized paperback and very light, the 810 would make the bits go in and out of memory, routing them properly between all my new toys, which clung to my computer like mandibles on a sedentary, deeply satisfied bug sucking in data.

My dad bought the 810 for me for less than $100. Without realizing it he'd set me up. At school Roger gave me phone numbers—Pat's, Aladdin, Pirate's Cove—a floppy marked CHAMELEON CRT TERMINAL EMULATOR, and a little booklet that I photocopied in the library. The book was a manual that explained how to set up CHAMELEON so that my Atari could call another computer and morph into a dumb terminal through which I could control the computer on the other end of the line.

These were pirate boards, bulletin board systems (BBSes) set up by kids with an Atari, an Apple, or a modem and a phone line. Some worked only at night after parents were asleep, when they didn't mind their phone line getting used. Others were run by older people whose phone lines were up all the time. Whatever the hours, getting hooked was tough. One line. Cracked software waiting for download. Busy signals. I had a speed dialer. It dialed numbers fast, over and over. The phone line went into the modem, the modem went into the 810, the 810 went into my Atari, and the Atari went into my brain. There at my desk, the little overhead light on, my TV screen glowing, speed dialer dialing, my Percom dual-drive unit read-writing, I was going under; I pretended I was in a sub about to dive. *Modem on,* I would say in my head in an authoritative, mission-control voice. *Check,* another voice would respond. *Full Duplex. Check. Parity Off. Check. Flow Control ^S/^Q.*

Check. Dial. Down, down I would go, the tones signaling the mission was on.

```
CARRIER 300
CONNECTED
```

And a world of choices would appear.

```
U)pload files.
D)ownload files.
L)ist files.
H)elp.
Q)uit.
O)ther users.
```

In in in. Every BBS had a menu like this one. Behind it were byteloads of cracked software. Games. Free. If you had new cracked games not on the system you were expected to upload them. Have one, share one. We were a collective, out for ourselves yet working together. It didn't seem like stealing, although I knew that the people who wrote these games felt it was. Copy-protected software violated our code. Programs were meant to be shared, and many had been altered and improved by kids. Looking at the source code, I learned how the games worked; I could alter the lines, teaching myself through trial and error how to program—which was as much a game as the game itself. Games that came compiled in binary were almost as bad as copy-protected games. As a binary file, there was no code to read, just incomprehensible 0s and 1s. A binary file could teach you nothing. Nor could it be changed. It just *was*.

I don't know how many kids were out there with me over the wires those nights, but we must have been a minority within a minority. Whatever the number, there were enough to support dozens of BBSes within my area, which was measured by area code: 212, 516, 201, 914. Any farther out than that—to 415, 213, 202, for example—and the long-distance charges would anger my dad. I heard they had good boards in 415. San Francisco. Too expensive. Some kids talked about 011–44, UK, where they had other computers called BBC Micro and Acorn. Excellent! We could go anywhere, in a world so new it had no name, a world we'd someday call cyberspace.

While those faraway places would have to wait, there were adventures to be had close to home. I had an idea. With the chip boom came gadgets, electronics stores, and chains of electronics stores. One of these chains had a store in Midtown a few blocks from my mom's house that was gadget heaven. Calculator watches. Rubik's Cube. Rubik's Pyramid. Rubik's Sphere. Handheld games. Calculator credit cards. Glass balls with lightning in them, static that leapt to touch your fingers when you touched the ball. Stereos with great lights, red dots, green dots, blue numbers, always flashing. Computers. Software. Games. It was a good place to go and play. I was there one afternoon, a few weeks after I first went online.

Next to the cash register was a computer, an IBM-PC, and I started talking to a salesman about it. He seemed to be a few years older than me, with a wispy mustache and pock-marked skin. I asked him why they were using an IBM-PC. The PC had been around for a year now, but I hadn't had much time on the IBM. It cost $3,000. He said it was good for accounting and keeping track of inventory. They call in at night to see what we sold, he said, referring to the central office. They can see what store is selling which items and order more through the computer. I looked at the PC. Next to it was a modem. A Hayes Smartmodem, like mine. I looked again. A number was taped to the top of the modem. Seven digits. A local phone number. A classic example of poor security. The guy kept talking. I memorized the number.

At home that night I waited for my dad and Joanna to leave for dinner. I kept going into their room to see how far along they were at getting ready. "Aren't you going to be late?" I asked, exasperated but trying not to show it. My dad was putting on his shoes. I went to my room. A few minutes later I came out. Joanna was putting on earrings. *Will they ever leave?* I felt like screaming. "There's food in the fridge," my dad said as they headed to the door. "Okay. Bye," I said, following them.

The telephone rang. My dad could stay on the phone for hours. *Not the phone, not the phone!* I screamed silently. "I'll answer it," I called out quickly. I ran to the phone and picked it up. It was my grandmother. "Grandma!" I said loudly, looking up at my dad and waving him on with my hand, letting him know I'd talk to her and they could go. I listened to my grandmother with one ear as the front door swung shut and closed. It was time. "Grandma," I said, "I have a lot of homework. I have to go."

I am going to break into a computer tonight, I said to myself. I went into the kitchen, got myself a big glass of water, and took it to my room and sat down at the Atari, booting up systems, speaking out loud, a submarine commander about to slip into bottomless cyberspace. *Power on. Check. Modem on. Check. Emulator booting. Check. Full Duplex. Check. Parity Off. Check. Parity Even. Check. Baud Rate 1,200. All the way,* I thought. *1,200. Check. Dialing, dialing* I said to myself, picturing the sub going under, into the blue screen.

Screeching bits.

CARRIER 1200
CONNECTED

We're in, at 1,200. The name of the chain came up. A menu screen. No password. I was right in. No password because who else would call? In 1982 who else was cracking computers? It wasn't a big thing yet. *War Games* was still a year away. I stared at the menu, reading each choice over a few times. Look at inventory? Search for a product? Order a product? Enter a customer's address? Read a customer's address? Download data? Get help? The screen was everything now. I was out of my body. Let's look at customers. Alphabetical or by Zip? Zip. I was 10021. I looked through my neighborhood and didn't recognize any names. Inventory: Alphabetical, Stock Number, Type. Alphabetical. A list of products came up: APPLE II PLUS, COMMODORE 64. I recognized some games—ULTIMA, CHOPLIFTER, RUBIK'S C, RUBIK CHAIN. Chain? The key chains. The store sold tiny Rubik's Cubes for $5, less than an inch on a side, minipuzzles on key chains.

I could order these for the store. I pictured a delivery truck pulling up with crates of Rubik's Cube key chains. I looked at the menu. It had an order function. I pressed the button. PRODUCT CODE? The code was a number next to everything in the inventory list I'd looked at. I went back, got the list, found the key chain, and wrote down the number. PRODUCT CODE? A bead of sweat fell in my eye. I wiped it away and entered the code. QUANT? Another drop rolled off my head and fell into the keyboard. How many, how many? Did they come in boxes of 100 or was it 1 for 1? 1 for 100? 1 for 12? I typed 9, then 9 again, then 999. It wouldn't go any further: 99 was the most. Ninety-nine hundred or 99 key chains would arrive? A spasm went down my right arm. I hit ENTER. In. ARE YOU SURE Y/N? I was committed. Yes.

NO CARRIER

The line went down. Abruptly. I'd resurfaced. I looked around the room, dazed. I'd just ordered all this stuff, and then it cut me off. Did they know? Maybe they could trace my number? Someone, a SysOp, had been watching. It cut me off. I turned off the modem. What if they found me? *They wouldn't*, I thought, *no way*. But what if they did?

For a week after that night I was jumpy every time I went home. I kept waiting for the phone to ring and my dad to come in and say, "I just got a phone call. . . . " I didn't go on the wires for a while. I decided to go down to the store and have a look-see. Would they have lots of key chains in front? The next day I stood outside the store, afraid to go in. Mustering the courage, propelled by curiosity, I finally stepped inside. I sauntered over to the cash register where the bins of key chains were. I spotted the Rubik's key chains and moved in closer. More? The same? Just as I was trying to ascertain whether or not there seemed to be too many, I heard a voice behind me.

"Hello." I spun around, startling the salesman—the same one from before.

I looked at him. I had to pee. I had to leave. I wanted to know. It just popped out of my mouth. "You have a lot of these," I blurted, pointing to the key chains.

"They're five dollars," he said.

He seemed perfectly calm. Was he covering, or was he telling the truth? Had it worked? I wasn't sure. Maybe it had. But I could tell I wasn't in any trouble. That was the most important piece of information. I'd gotten in and gotten away with it. I knew the product code on the Rubik's Cube key chain! I knew the phone number. I could come back any time.

I don't know what would have happened had I been caught. Kids breaking into computers was still a new thing, and no one knew what to make of it at the time. It was a game, people could see that. I knew it was bad, but it was fun, like blowing up firecrackers or launching fireworks from the roof of my building. BBSes didn't have passwords; kids ran them, for kids. Few paid to go online. Information was everywhere. By 1983 there were 2,400 computer books in print. Magazines like *Creative Computing* had circulations of 250,000. Some, like *Byte* and *PC*, had 600-plus pages—fatter than *Vogue*. Children and teenagers were a substantial part of their readership. And computers sold and sold: 216,000 made it into homes in 1980 and by 1982 the number reached 2.3 million; by 1983 it was over 5 million.

My interest might well have plateaued at the limits of what a self-taught hacker and programmer could learn on a home computer. Yet I soon would discover someone who knew far more than I could imagine.

*   *   *

In the midst of this "computer revolution," as it was beginning to be called, I was one of the luckiest of the lucky. I didn't just have a computer at home, I had a special one at school. Few schools offered computer courses in 1982; mine was one of them. At Horace Mann on the third floor of Tillinghast Hall, high above the football field facing north was a magic room in which I would eventually spend more and more time seated at a terminal, controlling a DEC PDP–11/44. It was here that I came to understand computers deeply, and through them enter the future, one my grandparents never imagined and my parents only barely grasped.

Walking into the Horace Mann computer room one spring day in eighth grade, I picked up a request form for a computer account that had to be signed by my parents. That night at dinner I handed it to my dad and told him, "You have to sign this." I'd filled in everything—my name, my grade—all that was missing was his signature. He read it, stood up, and took a pen from the kitchen drawer and signed the bottom.

The next morning at my first free period I ran to the computer room and handed the sheet to Mr. Moran, the teacher in charge. "Come back tomorrow," he told me, putting the sheet in a little box beside his desk, "and your account will be ready."

Up until then I thought we knew everything.

Here was one adult who knew how to go further.

6

Dungeon

FEW STUDENTS AT HORACE MANN knew of the computer room on the third floor of Tillinghast Hall. In the spring of 1982 when I was in eighth grade, computer class was an elective and few students chose to attend. Older students cared less about computers than younger ones, and only half a dozen out of approximately 160 seniors had taken classes with Mr. Moran; a greater number of underclassmen had. Still, the number remained small: in 1982 a dozen freshmen were enrolled in computer classes. In my grade there was a tiny boomlet. At least 20 of us, from a class of 120, had signed up for computer accounts (approximately 40 more would join our grade the following fall as we entered high school). A year later this would change: all the seventh-graders would take at least one trimester of computer instruction, as mandated by the school. The administration had concluded that a complete education had to include some familiarity with computers. Ours was the last class to graduate without such a requirement.

In that last year of elective computer study the computer room was a semisecret place known to most kids only as a door, set between rows of red freshman lockers. This hidden quality reflected the place computers occupied in the school social hierarchy. Three years earlier computer study had been little more than an extracurricular activity, one notch above chess club or model United Nations. Among extracurriculars, computer programming generated no social cachet. Compared with A-list activities like being on the staff of the school newspaper, editing the yearbook, or becoming part of the student-run Governing Council, programming computers had no buzz. Because Horace Mann was known for academic excellence, extracurriculars were an essential part of a successful college application. As high school progressed each rising class competed fiercely for the highest post in their chosen extracurricular. It was understood that the troika of highest positions—student body president, editor-in-chief of the *Record*, and editor of the *Mannikin*—led all but surely to admission at Harvard, Yale, Princeton, or Stanford.

Although almost no one knew about it, the computer room had a similar position: PROMAN, which stood for Project Manager, given to a few students every year by Mr. Moran. Every recipient went on to attend Harvard, Yale, Princeton, or MIT. In the vernacular of the computer room the PROMAN was referred to as "Super User." This was not an academic distinction (although anyone who earned it typically received straight A's in computer class) but a mark of responsibility. Super Users were system administrators. They ran the room as much as the teacher. A Super User was expected to be present whenever Mr. Moran was out of the room. Super Users didn't just baby-sit. They installed and upgraded new programs that were made available to everyone. They also wrote software, creating applications that otherwise wouldn't exist, which in part reflected necessity. In 1982, when computer education was then in its infancy, student-crafted software was essential to the curriculum. Schools nationwide hadn't yet begun to invest in computer courses, and few companies provided software environments for teaching below the university level.

In his quiet way Mr. Moran had created something that existed nowhere else at Horace Mann: student teachers. There was no boundary between learning in class from Mr. Moran and learning out of class from Super Users and younger students who might one day become Super Users. Between officially scheduled classes anyone with a computer account could come into the room and log in to the system, the Programmed Data Processor (PDP) model 11/44 built by Digital Equipment Corporation whose components, system, and cooling units had cost $200,000. Through organic evolution, trial and error, Mr. Moran created an open system, both in the way the computer functioned and the social fabric of the computer room. The machine and students existed in symbiosis, each part of the other.

Without the student system administrators writing programs, updating software, managing younger students and answering their questions, Mr. Moran would have been incapable of being teacher, guide, administrator, and sometimes policeman. More important, centralized control by one teacher was antithetical to the ethic of gleeful exploration and discovery fostered by access to computers. Mr. Moran recognized this all-important principle—of decentralization and communal ownership of the computer system—by empowering kids to strive for total access. The Super User had no restrictions: with the title came the right to access any other student's account, including that of other Super Users. A Super User could, in principle, crash the system, delete every file, or snoop anywhere. This was not a sophisticated ploy at divide and conquer or reverse psychology but rather a reflection that for kids to be well-educated, responsible citizens in a Digital Age knowing how computers work is not enough. Education is not complete without a genuine understanding of moral and ethical questions raised by information technology. Who owns software? Where does someone's electronic property or territory begin and end? At what point do shared systems become public? In 1982, when I entered the computer room, its organization seemed a harbinger of how children would be taught to use computers; little did I realize how soon it would become an aberration, an experiment abandoned once personal computers became too powerful for children to understand.

That spring of eighth grade, what first called me to the room was the thrill of adventure. It was too late for me to take a computer course—that would have to wait until ninth grade—so I went there to play. During a free period one balmy afternoon in April, my computer account activated, I discovered something exciting. Seated at a terminal at the far end of the room next to a wall of windows overlooking the football field, I found a game.

```
You are in an open field west of a big
white house with a boarded front door.
There is a small mailbox here.
>
```

Through lead-paned windows streamed shimmering rays that bounced off wooden floors and covered my hands with warmth as languid, swirling motes of dust flashed and floated in beams of light. I sat at the main table that ran the length of the room with a half dozen terminals on either side. The head of the table abutted Mr. Moran's desk; the other end, near where I sat, was pressed up against the wall of windows. Behind Mr. Moran's desk was the blackboard. Taped above were long printouts of text art, or ASCII art (pronounced "ass-key" by those in the know), which were images made of letters that from a distance looked like Mona Lisa and Snoopy. A door to the right of the blackboard led to the hall outside. A second door, in the wall on the right, led to the system room. Inside, visible through panes of glass, was the PDP and a table that served as Mr. Moran's private office. The system room was off limits to all but Super Users and Mr. Moran, who from time to time might invite a student inside. The opposite wall on the left had another long table, filled with electronic gear I didn't recognize, apart from the printer and an Apple II, which few kids ever used.

```
>GO EAST
The door is locked, and there is
evidently no key.
>
```

The sharp sound of a whistle outside, baseball coach yelling, a missed play. The athletes below played their games. We played ours. After school some practiced sports, many went home, and a few sat here, escaping into another world. Mr. Moran graded programs at his desk. He was a big man with a red beard, red hair, and heavy forearms like a lumberjack's. His eyes were luminous gray-blue and he always wore silver glasses, which softened his appearance and gave him the look of one who noticed everything. He existed in my mind as a descendant of Vikings, a benevolent warrior at the helm of our ship, guiding the room through waters of knowledge. Behind him the blackboard, scuffed with yellow chalk, revealed the wonders of logic gate design, control structures, database design, hexadecimal arithmetic. On his desk a sign read TO ERR IS HUMAN. TO REALLY FOUL THINGS UP YOU NEED A COMPUTER.

Silence, except for the soft hum of the PDP's drive heads read-writing in the system room and the erratic clicking of keys.

```
>GO NORTH
You are facing the north side of a white
house. There is no door here, and all
the windows are barred.
>
```

This was "up the hill," the top of it, the highest point of the school campus and for me the summit of knowledge. From the computer room, if I looked up from my terminal I could see a great expanse of trees stretching to the northwest, with the Soviet diplomatic compound a few miles away, satellite dishes on the roof, antennas rising higher still. I was LOGIN-ID 186,19, User 19, class 186. Class of '86, the ninth eighth-grader out of 120 to ask for a computer account.

```
>GO SOUTH
The windows are all barred.
>
>GO EAST
You are behind the white house. In one
corner of the house there is a window
which is slightly ajar.
>
```

In 1982 the computer room was three years old. Before 1980 there had only been a closet-sized computer club

room. Created in the early 1970s by a first generation of
Horace Mann students with math and engineering interests,
it had been located where the system room would later be
(back then the computer room was a math classroom). In
the club room students would gather around a teletype machine, which sat on a pedastal and housed a combined keyboard and printer. It was a boy's room; girls were excluded
by its clannishness. Each class of boys competed and collaborated as in a complex game of Dungeons and Dragons,
passing on knowledge to the new kids who came every
year. Working with girls that way required a level of intimacy and comfort between the sexes few teenagers were
capable of. The lure of the tiny room was its clubhouse feel,
a place for the intellectually curious to play together.

```
>OPEN WINDOW
With great effort, you open the window
far enough to allow passage.
>
```

The first generation of computer room boys had gotten
their hands on a teletype, the input and output of which appeared as printed text on perforated rolls of paper that fed
into the machine, and a modem. The teletype and modem
connected the students to mainframe computers. Wherever
someone knew of a password and login name, the teletype
could reach out through the modem and give that person
access to a computer far away. With older siblings at colleges, some of which were equipped with new, campuswide
computer systems, getting access to these computers wasn't
too difficult. For those who were into computers information spread quickly—which mainframe to call when. A little
club had formed and it contained the spirit that in 1982 was

still palpable in the new computer room: a desire to explore new systems and go where no one had gone before.

In the late 1970s the school administration, prodded by the excitement of the boys in the computer club and the dramatic arrival of home computers, decided to offer a computer class that in its first incarnation used monitors to connect with a PDP at nearby Riverdale High School. Riverdale was Horace Mann's arch rival on the athletic fields and a rival in the quest to lure prospective students. For Horace Mann to use Riverdale's computer system at first seemed prudent. While home computers may have seemed to some a passing fad, when it became apparent that they were here to stay Horace Mann's trustees began to reconsider: Should the school perhaps offer an expanded computer curriculum and acquire a computer of its own?

```
>ENTER
You are in the kitchen of the white
house. A table seems to have been used
recently for the preparation of food. A
passage leads to the west, and a dark
staircase can be seen leading upward. To
the east is a small window, which is
open.
On the table is an elongated brown sack,
smelling of hot peppers.
A clear glass bottle is here.
The glass bottle contains:
A quantity of water.
>
```

When Horace Mann connected to Riverdale in 1978 the administration hired Mr. Moran to offer courses in programming, computer architecture, and Boolean logic. It was Mr. Moran who produced a proposal, in 1979, for Horace Mann to acquire a PDP of its own. He wrote it with the help of the computer club's members, some of whom went with him to the school's board of trustees to petition for the new curriculum. At the time Mr. Moran argued that this generation of children and those who would follow were entering a new time, the Age of the Computer, he told them, and that any well-rounded education had to include computer skills. He proposed building a computer science curriculum that began with the basics—BASIC programming—and moved upward into the exquisite realms of bubble sorts, binary sorts, searching algorithms, hashing searches, linked lists, sparse arrays, full arrays, recursion, trees, sets stacks, queues, lists and matrixes, and languages LISP, FORTH, FORTRAN, COBOL, Ada, C, ASSEMBLER, Pascal. He would introduce students to corporeal architecture: AND-OR-NOT gates, the central processor, I/O, and peripherals such as printers and modems.

The board responded to Mr. Moran's proposal favorably, swayed in part no doubt by the sight of well-dressed computer club members in their best jackets and ties offering testimonials. In spring 1980, the trimester before I entered Horace Mann, the PDP–11/44 arrived, hauled up the broad steps of Tillinghast to the third floor and placed in the special room with its own air-conditioning, telephone lines, and extra power circuits.

```
>TAKE ALL
brown sack:
Taken.
glass bottle:
Taken.
>
```

Three years later, for those of us with home computers the PDP offered a way to accelerate our progress by distilling eclectic knowledge gained from magazines and hacking programs into a set of principles. In eighth grade I knew by instinct how programs worked, but not why. *Compute!* and *Antic* magazines offered us "how to" solutions in the form of long printouts of programs with which we could experiment at home, learning by doing. The computer room added to this hands-on experience by giving us an understanding of classes of solutions, of archetypes. It gave us access to models of thinking that, while we did not know it then, existed behind every line of code I copied out of *Compute!* In this new environment I took the first steps out of our vernacular, where programming tricks were passed as lore from kid to kid, into the formal realm of programming theory. Here philosophy came alive in the form of our games, our programs, the operating system, and computer languages housed on the PDP's metallic disk drives.

Our computer education was organized according to the same principle that grounded my liberal arts education at Horace Mann: education meant learning how to think. In history and English classes we learned that what matters is asking the right question, not memorizing the correct answer. Facts are important because they serve as proof for the ideas we craft. This ideal is at the heart of humanities instruction, and modern-day teaching is rooted in science. Scientists gave us the techniques we use in English and history. Our teachers' emphasis on building models is rooted in the scientific method, whereby a theoretical model is tested by means of experiments. The model is then altered to fit the real-world outcome, based on the results.

```
>GO WEST
You are in the living room. There is a
door to the east. To the west is a
wooden door with strange gothic
lettering, which appears to be nailed
shut.
In the center of the room is a large
oriental rug.
There is a trophy case here. On hooks
above the mantelpiece hangs an elvish
sword of great antiquity.
A battery-powered brass lantern is on
the trophy case.
>
```

The scientific method is iterative—that is, conclusion built on proven conclusion, bit by bit, and iterative thinking is one of several core skills we acquired day after day with Mr. Moran. The scientific method gave us a model-making tool kit, a set of design rules for creating ways of thinking about the world and then testing our ideas. In our liberal arts classes we were given tools to interpret texts, then were expected to formulate coherent arguments of our own. This, more than the ability to recite Shakespeare's sonnets, was the literacy our school promised our parents we would acquire. Dexterity with tools for thinking rather than the mechanical ability to recite Homer signified the formation of a well-educated mind. What we had in the computer room was the extension of this literacy to the digital frontier. We were given tools to analyze and deconstruct computer programs and thus acquired the ability to understand computer-mediated systems.

In the computer classroom we were being prepared for adulthood in a new world where information about what was happening at any second in time went from scarce to abundant. Without preparation such plenitude could become a curse, experienced as information overload. For the digitally literate, however, this overflow of data could serve as building blocks that turned new information into new knowledge. We were being prepared to fashion meaningful interpretations from the flood of raw data brought on by cheap computers. At the core of these emergent systems, from ATARI-DOS to credit reports, medical records, air-traffic control, video games, word processors, font design software—all the infinite permutations of what a digital computer can create—were certain principles, a way of thinking. Paramount was the sense that everything in the world was based on interlocking systems, and systems of systems. This was reflected in the architecture of computers and software.

```
>OPEN CASE
Opened.
>GET LANTERN
Taken.
>GET SWORD
Taken.
>MOVE RUG
With a great effort, the rug is moved to
one side of the room. With the rug
moved, the dusty cover of a closed trap
door appears.
>
```

Once you understand the inner workings of a computer, the atomic world of bits and the microscopic world of coding, you can understand the macroscopic world of software and the networking of networks of people. Boolean logic and principles of computer programming, like recursion, first emerge as the innermost building block inside a computer and then, remarkably, reveal themselves again and again as you widen perspective, reemerging in the function of finished programs, networks of programs, and networks of computer networks. What worked in explaining computers lent itself as a model for looking at natural systems like weather, abstract systems like markets, and social systems like politics.

I joined a community of people who thought the same way I did. Our idol was Mr. Spock from *Star Trek*. Spock, the alien from planet Vulcan with pointed ears and eyebrows, could not express emotions. He solved problems through the exercise of logic. The human crew of the starship *Enterprise* would turn to Spock for advice, relying on his ability to solve problems through reason. When a life-or-death situation required immediate attention he often saved the crew with a disciplined, model-based step-by-step analysis of the problem. This much, we thought, was the way our computers functioned, a way that worked best if you believed the world was a function of ever more complex systems, each grounded in fundamental principles that held true, whatever the scale. The mystery rested in untangling complexity, and doing so required the fortitude of the puzzle-solving explorer.

```
>OPEN TRAP DOOR
The door reluctantly opens to reveal a
rickety staircase descending into
darkness.
>GO DOWN
```

```
It is pitch dark. You are likely to be
eaten by a grue.
Your sword is glowing with a faint blue
glow.
>
```

   That first day, LOGIN-ID 186,19 gave me certain rights in
our shared community. I had the right to 20K of disk space, a
sliver of our shared RA80 and RK70 disk drives that spun at
100 miles an hour behind the glass wall, seeking data in 49
milliseconds, sifting through 177 megabytes of storage that
contained every program, every term paper, every little file
we created. I had the right to command 64K out of the titanic
1.5 megabytes of RAM that formed the dynamic memory of
our PDP. These were the rights every user began with. Of
course, with age and experience greater rights could be had.
The 20K of disk storage could grow to hundreds, perhaps
thousands, of K. The 64K ceiling could be breached, 32K
slices of memory annexed, fused to create a greater stack.
Super Users held the keys. They managed our system, allo-
cating these scarce resources. Helping them were a set of elite
younger students known as Senior Level Users, or SLUs. SLUs
were anointed at the end of their first year of computer study,
allowing them to advance to higher levels of programming
with sufficient disk space to experiment with larger programs.

```
>LIGHT LAMP
The lamp is now on.
You are in a dark and damp cellar with a
narrow passageway leading east, and a
crawlway to the south. To the west is
the bottom of a steep metal ramp which
is unclimbable.
```

```
The door crashes shut, and you hear
someone barring it.
>
```

Super Users and the SLUs were heirs to the ethic of the original computer club, when kids maintained the machines and decided what projects to pursue. Programming is learned by trial and error through direct interaction with the computer, and the most efficient way to manage this process is by enabling students to control their own time, to choose their own path to their goal. If playing games led to an intuitive understanding of directory trees, command structures, and the step-by-step nature of digital machines, then playing games was one route to the end. If hacking apart other people's programs to see how they worked and then changing the program to make it do something else served to teach database structures, communication with the central processor, and flow control, then hacking was an appropriate path to the end.

Some goals required all the students to participate, each taking on a part of the task. For instance, the PDP used an operating system called RSTS/E version 8. When updates to version 8 from Digital Equipment arrived on big plastic magnetic tapes, the SLUs worked together to bring the new version online, taking collective responsibility to update the system, which often demanded bits of programming that required debugging. At stake, in a worst case, was the entire system.

```
>GO EAST
You are in a small room with passages
off in all directions. Bloodstains and
deep scratches (perhaps made by an axe)
mar the walls.
```

```
A nasty-looking troll, brandishing a
bloody axe, blocks all passages out of
the room.
Your sword has begun to glow very
brightly.
>
```

All student files, games, programs—everything—could be permanently erased by an SLU with a few keystrokes. But such fears were the fears of another place. In our room no one worried that an SLU would erase the system. Students were trusted. Mr. Moran knew such a world was possible. He designed it, the way an architect creates a building knowing that one day it will be done and that the owners will take control. With the delicate touch of an impresario in the wings, Mr. Moran let us run the show, yet the show ran within the subtle parameters of the space Mr. Moran designed. As a willful act of co-creation and self-organization, teacher and students collaborated on a shared journey.

```
>HIT TROLL WITH SWORD
You charge, but the troll jumps nimbly
aside.
The troll swings. The blade turns on
your armor but crashes broadside into
your head.
>
```

That May afternoon a few kids worked at terminals coding in BASIC, while some of the older ones worked in Pascal, FORTRAN, and C. A few hacked in ASSEMBLER. Others played games: CRASH, LUNAR, SUMER, DND, TREK, DUNGN. Six letters maximum in a name, with three-letter extensions: .BAS, .PAS, .TXT. I'd typed RUN [GAMES]DUNGEO that afternoon. I had permission. Dungeon was a big game, one of the biggest. Only one person could play Dungeon at a time, and then only if Mr. Moran allowed it, if it was a free period and other kids didn't need the processing power. Processor time, disk space—the assets we shared.

The day I discovered Dungeon I decided the computer room was the best place at school. There was so much to discover here, and each level would be just as much fun to explore as the one before it. It would never end. The "it" was a feeling, more than anything else, of excitement that the unknown could be known, and that discovery brought joy. Dungeon alone would take weeks to solve, and no session was ever the same. Unlike video games, Dungeon exhibited uncanny flashes of sentience, of randomness, personality. Whenever my free periods and the computer room's open times coincided, I was there. I would stand outside the door in the hall, waiting for a computer class to end, peering though the window at the kids inside, my heart racing. I had to get in first, log in, and boot up Dungeon. The first up with Dungeon got to play; anyone else had to wait, otherwise the computer's memory would be overwhelmed by versions of the game running simultaneously. Sometimes the race to log in made Mr. Moran mad. We never wanted to make him mad. Our running to a terminal jarred the serenity of the room and seemed vulgar. We had to walk to the terminals, which we did as quickly as we could. Loud shouting or complaining that someone had been playing too long was also bad. We had to be adult about it and wait patiently.

```
>RUN EAST
The troll fends you off with a menacing
gesture.
```

```
The troll hits you with a glancing blow,
and you are momentarily stunned.
>
```

If Dungeon was taken, I would dig through the [GAMES] directory and find something else, something smaller, that I could play. These smaller games taught me subtle things. One afternoon I found a game called SUMER, as in "Sumerian." The computer took on the role of Grand Vizier, with me as Emperor of the Babylonian kingdom. At the start of each cycle, or year, I had to decide how much money to spend planting food and how much money to collect in taxes. The computer then made its calculations and announced the results. A good year meant having surplus crops and a healthy population; a bad year came around when people starved to death and revolted, ending the game. The game turned abstract concepts of inflation and exponential growth into concrete consequences. Too much food, too much generosity, might lead to a population explosion and dry years ahead. Stinginess could suddenly and dramatically implode into famine, wiping out farmers and guaranteeing that the next year would be even grimmer. Getting it right was hard. Sort of like life, I imagined.

Other games, like CRASH, gave me a sense of "leverage," the idea that with a little force you could move a huge object. In this game leverage meant borrowing money, investing it, and using the profit to pay back the loan, thereby leveraging my investment. Of course a wrong bet would immediately wipe me out. These games took very little disk space, and were like cookies—little brain snacks to nibble on. In all these games I sat in the same place, as an omniscient outsider with total power over the hermetic universe before me. A system for me to analyze. After years of playing video games the dense, text-based universe of the Horace Mann computer system lured me with its promise of knowledge, power, and tantalizing possibility that one day I too could understand how these programs worked, and write my own.

```
>HIT TROLL WITH SWORD
You are still recovering from the last
blow, so your attack is ineffective.
The troll swings his axe, and it nicks
your arm as you dodge.
>HIT TROLL WITH SWORD
A mighty blow, but it misses the troll
by a mile.
>
```

We had no color monitors, no video games to play. Everything here was text, blocky white letters on black screens. Yet the words were seductive because they revealed a hidden order behind everything. My Atari would boot up from a disk, hauling a game fully formed into memory; on the PDP, however, we existed as a tribe, each of us allocated a small amount of space, on our own, yet aware of whoever else also was online. I could look up and see the faces or I could type USERS and get a list of who was logged in and how much of our shared processor time they were consuming. There was an order to the directories, which nested, one inside the next, like Russian dolls. There was an order to the programs, to the printouts in BASIC, FORTRAN, and COBOL. There was an order to everything, and I sensed this hunger to find out what it was. Where did the order end, or begin? What was its source?

```
>HIT TROLL WITH SWORD
The troll is momentarily disoriented and
can't fight back.
>HIT TROLL WITH SWORD
The unconscious troll cannot defend
himself: he dies.
Almost as soon as the troll breathes his
last, a cloud of sinister black smoke
envelops him, and when the fog lifts,
the carcass has disappeared.
Your sword is no longer glowing.
>SCORE
Your score is 35 [total of 585 points],
in 29 moves.
This gives you the rank of Amateur
Adventurer.
```

Dungeon was the second game, after Dungeons & Dragons with Jesse in fourth grade, that revealed how computers worked. As their names indicate, the two games were related. Dungeon turned the computer into a Dungeon Master, running a solitary version of D&D with a lone human player—myself. The structure of Dungeon, however, and the imaginary world it created encapsulated the core principles of software: flow control, database architecture, control structures. The basic elements of software architecture and programming in Dungeon were there to be discovered indirectly.

Standing by the white house for the first time, two impulses took hold of me. I wanted to win the game, collect all the treasure, solve all the puzzles, know every room, sense the scope of this world; at the same time I was asking myself how this worked. Why was it that when I wrote "open window" the computer understood me and responded? How did the computer know which objects could be picked up? I could take the bottle of water on the table, but I could not take the table itself. How did the computer know what time it was? Or arrange for me to encounter a nasty thief who kept stealing my treasure, or keep track of how many blows we traded, and when it was time for me, or him, to die? Later in the game I was able to get into a boat and float down a river. How did the computer do that, put me in a moving object that moved through other "rooms"? There were so many possible combinations of words—too many to list in one long file with canned answers, so I knew that somehow the computer had a kind of intelligence; it was able to understand grammar, usually in the form of two-word verb–noun combinations such as "take sword," and the meaning of certain words.

I understood that something called a "parser" did this, because I'd read the Help file that came with the game. The parser was the part of Dungeon that accepted my typewritten commands and then scanned its lexicon of words to interpret the meaning of my command and respond with a new description, one affected by my location in the game, what I was carrying, what else might be in the room, my health, and the existence of a hostile character opposing me. Since all these variables were changeable the computer's response often varied according to the specifics of the situation. A lot of what I did in Dungeon became curiosity fulfilling, along the lines of wondering what the range of the computer's intelligence might be. What might happen if I wrote "take the troll to dinner," or "fuck you," or "do you believe in God?" (In order, the responses are "I can't see one here," "I don't understand that," and "I don't understand that.")

I now know that Dungeon works using a set of basic principles, a design or technique with two fundamental parts.

The first part is the data: the information about all the rooms in the game and their contents—the dungeon. The second part is the control structure, or rules that govern the dungeon. Like life which is composed of space and time, Dungeon shared the same model: space was the data, time was the control structure. Separated, both are meaningless; combined, they produce a world. The control structure looks at the dungeon and what you do at any given moment, evaluates the circumstances, and responds according to the rules it lives by. So for instance, when I wrote "kill troll with sword" the control structure looked at the data, saw that I was in a room—in this case "a small room with passages off in all directions. Bloodstains and deep scratches (perhaps made by an axe) mar the walls"—and that an object known as a troll was present in the room with me. When a "nasty-looking troll, brandishing a bloody axe, blocks all passages out of the room" the computer then made a decision—based on rules governing the object called "troll"—producing this:

```
The troll swings his axe, but it misses.
>
```

At the > I had a number of choices. I could write "go east" and run away from the troll, and the control structure looking at the data according to its rules would understand that I wanted to move east, and I would see:

```
The troll fends you off with a menacing
gesture.
The axe crashes against the rock,
throwing sparks.
>
```

I was trapped, unable to escape this fight. I had one weapon—a sword—that might be strong enough to defeat a troll wielding an axe. This introduced another concept: the control structure understands that certain kinds of data have different values. A sword is worth something different than, say, a hand. That difference translates into another kind of data: the "health" of the troll, or me. Hitting the troll with my hand would subtract less from his overall health than hitting him with a sword. Likewise, his axe would do me a lot of damage with just one hit, lowering my health and bringing me closer to death. I could know my health by typing "diagnose."

```
You have a light wound.
You will be cured after 29 moves.
You can be killed by one more light
wound.
```

Before I can type in another command the computer tells me:

```
The troll's axe bashes your skull.
It appears that the last blow was too
much for you. I am afraid that you are
dead.
```

Time passed in this game, even when you did nothing. This was an exciting concept because it created a feeling of tension and realism—this world turned without me. What made this possible is a set of three concepts governing the control structure. I later learned that these were called *subroutines*; I think of them as modules, each of which has a task to perform, producing a result that is passed on to the next module. The first is the parser. Once the parser takes my command and translates it, the command is formatted in an internal language and passed on to the second module, the command executor. The executor then alters the world according to my command—for instance, moving an object out of a room and into my possession when I write "take bottle." The final module is independent of me; it creates "events" according to the passage of time and the state of the dungeon. The "event module," after I pick up the water, might suddenly announce that a "seedy looking gentleman carrying a large bag" just came into the room. This is the thief, and he will try to steal whatever accumulated treasure I may be carrying.

The thief is a special module unto itself, what programmers call a *daemon* or agent. These are mobile subroutines (programs) that can move around within a single program or between separate programs (over a computer network). Today programs called "daemons" control the flow of e-mail through the Internet. In Dungeon this combination of rooms, objects, commands, daemons, and time are the same ingredients in most software. For instance, a word processor has data—in this case, information describing certain commands—that are like rooms. The option that allows you to select from a list of type fonts is the "font room," and the option that allows you to print is the "print room." These "rooms" are connected by a set of commands governed by the location of a computer mouse, which in Dungeon is like typing "go east." Your document is an object, like my bottle of water. The daemon might be the "printing" subroutine that is invoked after you've chosen to print your document. This daemon is the same in your spreadsheet program as well as in your word processor, a function of its mobility and use in multiple programs. The time function comes up when the computer warns you that you ought to save your document again, since it hasn't been saved in a while.

Long before I learned any of this in my computer classes Dungeon gave me an instinct for the feel of programs and operating systems—how they worked and how to design them. Dungeon also showed me that the computer could create anything I wanted, as long as I knew the commands, in their proper sequences. I understood that somewhere someone had written this program, thought out these puzzles, and put it all together in a piece of software I was now using. Dungeon offered me the seductive possibility of eventually building my own world, with its rules, parameters, and meaning. It seemed a natural enough idea to build a kind of fantasy world that other people could share. At the age of fourteen I was still spending a lot of my time daydreaming, fantasizing, running stories in my head. Dungeon was a tactile, physical extension of that impulse, but unlike internal dialogue, it could be shared.

Like Dungeon, the operating system on our school computer, RSTS/E (which stood for Resource Sharing Time Sharing Extended, pronounced "ristus") had its own logic, commands and levels, puzzles and secrets. With tenacious exploring, these could be found, gathered, and tallied. In Dungeon treasure had to be brought back upstairs, to the white house, and placed inside the trophy case; in RSTS (and the programming languages that followed) the treasure was less well defined, less tangible, yet more real. The treasure was knowledge, to know something deeply and well, to experience a world and move one step beyond—to learn how it is built and then set out to build it further.

*   *   *

I'd found another home. In the place behind the screen, in the realm of metaphor and words, the bits of the PDP became real, tangible. That spring, as I lustfully explored the PDP, impatient to know everything, I decided to enroll in my first computer class. That class was the first step toward the recognition we aspired to, the mantle of SLU, and for some of us, the title of Super User. I wanted that—to have the skills to be chosen, to be Super User. Three years of learning stood before me: three years of computer class, of programming, of growing up. Three years of learning, of apprenticeship to the mind of the computer—and to the demands of Mr. Moran. I wanted to be great at something finally, and I could, I thought, be great at this.

7

Super Users

I STOOD OUTSIDE THE DOOR to the computer room, impatient to get in. The halls of Tillinghast were stiflingly hot. It was the first week of school and Indian Summer had hit New York, humid air rising up the building and pooling along the top floor. The only air conditioner was on the other side of the door in the system room, that privileged vestibule. I'd waited all summer for the start of ninth grade, I'd longed for Computer I to begin, and now that I was on the threshold I had to wait.

Through the window in the door I could see the kids from the previous class sitting in two orderly rows, alongside the long table, terminals on. Mr. Moran was running late; he'd gone over and was assigning homework. I'd never wanted to be in a class this bad. Fuck! Come on! Kids shoved past me. Classes were switching, and hundreds of students filled the hall. My bookbag pressed down on my T-shirt and the sweat from my back had soaked through. The bag felt soggy. A voice behind me said "hey," and I turned around. Scott stood there, his hair matted on his forehead, clinging to his glasses. Behind me the rest of our class assembled. We'd all chosen to take this class, and some of us would be together for four years, going through every level of computer courses until the very end, when some of us would be made Super User. Where our skills might take us beyond that—graduation, the real world—was impossible to imagine.

The door suddenly opened and kids poured out, spilling into the hall, dispersing in a rush to get to their next class. I was in the room first, and I hurried to the right side of the table. The four monitors on that side, closest to Mr. Moran's desk at the head of the table, were the game terminals. These were the only machines the PDP recognized as having access to the [GAMES] directory. I grabbed a seat, sat down, and typed HELLO. The terminal woke up. In a flash I entered my account number and password and moments later I was in [GAMES], commanding RUN DUNGEO. Instead of Dungeon I saw "game in use." Stymied. Blocked. *Someone else must be playing*, I thought. The game player had to be here, next to me, in my row, since this was the only row with [GAMES] access. I looked up. There he was. A new kid. Two terminals away, *his* character stood by the mailbox at the foot of the hill.

The new kid had curly red hair and gold-framed glasses. I'd never seen him before. It felt good being an old-timer. Looking at my classmates in the room, I was among the least nerdy—I'd kissed a few girls, done drugs, and wasn't good in math. Most of the kids in the room liked the sciences; my favorite class, outside computers, was history. Other than Scott and Aaron, my other friends at school weren't taking computer classes. For them my interest in computers seemed strange, something I did on my own time, in a room they never visited. For me the lure of the room wasn't only the experience of exploring, alone; it was the social dynamic as well, the feeling of belonging in a community. Were it not for the communal environment—the terminals connected to a time-shared computer, the after-class open times in the room where I could program with older kids—the room would have been too lonely a place.

Suddenly I recognized the new kid: Jeremy Bozza; people called him Boz. I'd run with him during tryouts for the cross-country running team. Mr. Beisinger, the coach, had sent the entire team, along with the freshmen who wanted to join, on a 2.5-mile race through the park down the hill. I'd seen Boz in the pack, as we spread over the field and up into the wooded hills, tapering to a long line, separated by speed. The best on the team, the seniors, had come in under fifteen minutes. Varsity times. Panting and feeling like puking the whole way, I'd crossed the finish line around 18:30, way behind the older kids and somewhere in the middle of the ninth-graders. As part of reinventing myself I'd decided to join a team. Owing to my poor vision, games without balls were best, and running had been the logical choice. Boz had come in behind me. Our times were good enough. Anyone under 21:00 minutes could get on the team. Slower than that and you belonged on the girls' team.

I turned back to my terminal and looked out over the room—there was Scott, and kids I knew from last year: Ian, Peter, Adam, Aaron, Roger, Misha. All boys. Mr. Moran pulled out the attendance sheet and called out our names, checking us off. He then leaned down to his terminal and typed something. I saw my screen twitch. BYE it said. We'd all been automatically logged out by Mr. Moran, who wanted us to pay attention to him and not our terminals. I settled into my plastic seat, the sweat off my back cooling, and smiled. We'd arrived, and my first computer class was beginning.

I was in awe of Mr. Moran. I wanted him to think I was a good programmer. I wanted him to like me. Taciturn and soft-spoken, Mr. Moran didn't say much, but when he did I listened. He knew everything about our system, which was so much bigger than my Atari. Mr. Moran was nothing like the other teachers, who tended to lecture at you, surprising you with pop quizzes. While Mr. Moran also lectured, what he taught us was not for memorizing. Instead of taking tests we wrote programs. Mr. Moran gave us a problem, and we had to write a program to solve it. For instance, one of our first assignments in Computer I was learning how to use IF-THEN statements, and he asked us to write a program that found the highest number in a series. We were expected to solve it after class, and helping each other was permitted. We could also ask Mr. Moran questions before submitting our program. Where other teachers sometimes felt like adversaries, here we were on the same team. In this way Mr. Moran was a lot like Mr. Beisinger, as much a coach as a teacher.

We worked on projects, we practiced, we came in during free periods to improve our skills, we shared information. Sometimes we competed against one another. Sometimes the competition got rough. Feuds broke out, skirmishes were fought behind Mr. Moran's back, on the read-write platters of the PDP, and in our programs. What drove our competitive urges was the will to power and to knowledge. We all knew that a chosen few would be supreme one day— as Super Users—and the road from ninth-grader to Super User was long and arduous. It wasn't enough to win Mr. Moran's favor; you had to be the best programmer. If you were serious about the room, rivalries sparked from the friction between us, singeing egos.

"Mr. Moran said the color plotter is finally coming," Boz told me during cross-country practice as we huffed our way through the woods with a half dozen other boys. We'd waited weeks for it. Built by Hewlett-Packard, the color plotter enabled us to draw elaborate geometric images in full color. Four pens, following the lines of mathematical equations in our programs, produced swirls, circles, radiating lines; whatever we could imagine and program could be drawn.

When the plotter was delivered the next day we descended on it, circling the machine on Mr. Moran's desk. That year's Super Users and SLUs—Amy Bruckman, Paul Haahr, and Paul Hilal—got first pass, hooking up the plotter to the computer. They installed the system patch that would let us control the plotter. Paul Haahr, a brilliant eleventh-grader who could solve polynomial equations in his head and next-in-line apparent for Super User, started off by drawing perfect concentric circles. He was the prototypical hacker: inquisitive, skeptical, and fearless of other people's opinions. Tall and skinny with long black hair, his hands were his most arresting feature, with thin fingers and long, perfectly manicured fingernails.

As we watched, Haahr produced sine waves that rotated about a central point, creating waves of solarlike rays. I was too inexperienced to understand math at that level, so I sat quietly and watched. Haahr tapped out equations and looped them, producing one hypnotic image after another. Without knowing, by watching Haahr I was indirectly learning calculus, two years ahead of schedule. I jotted down his equations, wondering if you turned that number into a negative, what would happen? If you made the loop go from 0 to 1 in increments of 1/100, how would that change the image? What about 0 to 10 in increments of 1/100? How about an exponential curve? We bombarded Haahr with questions, and he demonstrated the answers. We watched something called a "logarithm" in action. Paul talked about "derivatives." I was a C math student, but now, in this context, I loved math.

A day later I came into the room and found Boz hard at work with the plotter. I was beginning to see him as a rival.

"Look," he said as I walked over. "I figured out how to draw mazes." Sure enough, the plotter was drawing a maze. "It randomly generates the maze," Boz explained, "and then it figures out how to solve the maze." "How did you do that?" I asked. Boz said he figured it out on his own. I was suspicious. "Let me see the program," I said. He showed me his printout. It was complex, hard to follow, with nested loops and long equations and conditional statements. *He's good*, I thought, humbled. I asked for the printout. In keeping with the ethic of the room, Boz gave it to me.

Later that afternoon I was printing out mazes. Paul Haahr came in and saw what I was doing.

"Hey, that's good," he said. "Did you figure that out yourself?"

I was tempted to say I did. "No," I said, "Boz did. He taught me."

"Boz did?" Haahr grinned a funny grin. "That's funny. I showed Boz how to do it. Let me see the program."

I displayed the code onscreen. "That's my program," Haahr said.

From that point on Haahr never stopped teasing Boz. Boz, though, would fight back. It became a game. Boz had taken credit when credit wasn't his to take. I would have done the same thing. It was then that I noticed Boz tended to run headlong into trouble. I was far more diplomatic. He was overt; I was covert.

Despite the rivalry and infighting, at the heart of the room that year was a search for truth that I learned to cherish, a search I experienced nowhere else. In other classes the search for truth felt abstract, but ironically, here in the symbolic world of bits, truth took on corporeal form. Here truth actually did something—built a picture, ran a game, calculated a database. Truth felt authentic. The quest for it forged a sensation of intellectual honesty that I'd never known before. Each new program that we wrote from scratch could appear different from the others and still work. This kind of clarity mixed with creativity and expression didn't exist in other classes. My modern world history teacher, Mr. Donadio, was the one to decide whether or not my essay on Napoleon was good. His grade ruled and there was nothing I could do, even if I thought it was unfair. In programming, however, we could all see whether or not it was good, whether it was beautiful or ugly, excellent or lame. I felt pleasure when I programmed, pleasure in the making, and this newfound sense of mastery seeped outward, spreading into other classes until something in me shifted. I began to experience school differently.

Early in the first trimester I received my first grade, for a program I wrote in BASIC. My program was printed on big white-and-green computer paper with sprocket holes along the side. On the top of the page in Mr. Moran's writing were the words *very good* and A–. I'd never gotten an A– in anything before in all my life. Even my father was impressed. That night I went through the program with him, going down the lines and explaining the code. I had found something I was good at, a class where going beyond a C was possible. Something got through to me—the possibility that I could maybe get good grades in my other classes. That first trimester of ninth grade, I made a decision: I would get good grades too. I had joined an athletic team. I was no longer a new kid. And I'd gotten an A in something. This year was going to be a good one.

*   *   *

The computer room in a primitive sense was a forerunner of the Internet. When we were logged in each of our accounts appeared as homes that together made a town. And it was up to us to run the place, to be good citizens. Part of the challenge we faced as users of an open system was the challenge of the commons: it takes only one person to wreck the community's shared space. For those with the least investment in the open system, the new kids like me, the temptation to abuse the group's trust was strongest. Although we learned by exploring the system, some things were forbidden, and Mr. Moran took abuse of these rules very seriously. Breaking into the computer, trafficking in pirated software, tampering with other people's accounts were all dangerous offenses. I'd signed a paper warning us of the price we would pay for breaking the rules. It sounded serious:

> Anyone erasing, altering or accessing any private computer data files or records, system programs or other privileged information on the computer, or even attempting to put himself in a position to do so, will face expulsion from school.

A copy of these rules were posted on the wall in the computer classroom. The words *or even attempting to put himself in a position to do so* were emblazoned in my mind. It was as if even the mere thought of, of . . . the thought of . . . it was hard to say: *the thought of breaking into Mr. Moran's account*—was dangerous.

The thought first came to me one afternoon during a free period while I was working on a program for class. It was quiet; I was in the room with a few students and Amy, that year's Super User, who was chaperoning us in Mr. Moran's absence. Amy was the only girl programmer I ever knew at school who was unquestionably better than me. When Mr. Moran wasn't in the room Amy usually was, and when she was there her word was law. But most of the time there wasn't much to police.

That afternoon Mr. Moran came into the room from lunch. His terminal was running OCCUPY, as it always did during the day when he was not around. OCCUPY locked his terminal and flashed the time onscreen, and he had to type his password to unlock it. Mr. Moran went straight to his desk, put down his papers, and sat down in front of the locked terminal. I stood up and craned my neck to look at his hands when he typed in his password.

"Please look away," he said.

I turned red, fixed my eyes on my terminal, and listened to his fingers tapping on the keyboard. I wanted to look up so bad! But I kept my head down. When he was finished clicking I looked up. I don't think he knew that I'd gotten up to look at his password. Mr. Moran didn't seem angry, and I headed out of the room and went to the bathroom, as if I'd been planning to go there all along.

Breaking into Mr. Moran's account was difficult. I didn't know of any student who had ever accomplished this feat. Mr. Moran took precautions. The first ring of defense was his aura—this alone dissuaded most people—the way his silence seemed to indicate that he could sense our thoughts, detect our intentions. For the rest who were still curious and thought about breaking into his account, the second ring of defense was the threat of expulsion. Everyone signed that paper, and we all knew the consequences.

If these fail-safe measures failed, the next levels of defense were technical. Anyone attempting to sit down at Mr. Moran's terminal and crack the OCCUPY program would be seen. It was possible, though, to log directly into his account from another terminal. I knew that his account number was a coveted low number—1,0. The lower the number the more powerful the account, and any number with one digit was a Privileged Account, one with systemwide access. A one-digit account meant that you could go into any account, any part of the system, and do anything you wanted. It also meant that you could give anyone else a one-digit account. The power to propagate. That would be cool.

One day I came into the room when Mr. Moran was out to lunch and sat at the far end, next to the windows. The login screen patiently waited for me to type in my account number. I typed 1,0.

Password:

The computer waited for me to type it in. I didn't know Mr. Moran's password. I knew it was twenty characters or more, much longer than the six-character student passwords. It could be anything—"the quick brown fox," "Frodo lives," "xyzzy," or just gibberish, "fh#A.,,cTT8no;)54"—*I could just try one,* I thought. Suddenly I remembered. Shit! I hit Control-C, the command to "break" and interrupt the login program. BIGBRO was watching! BIGBRO. I'd forgotten about BIGBRO!

I looked up. Did anyone see me? A few kids typed nearby, quietly. Amy was in the system room working on a terminal, and she didn't seem to notice. I didn't want Mr. Moran to walk in and see me sitting there, so I quickly walked around the table to the other side of the room and sat down at another terminal.

BIGBRO, and its sibling, LILBRO, were constantly running. I logged in to my account and did a SYSTAT to check the overhead load on the CPU by job and got a list . . . BIGBRO was up all right. Listed at the top. Damn. I had no idea what BIGBRO or LILBRO did. But one of them was always running. I stared at the statistics. BIGBRO was consuming a small percentage of our total processing power. The other programs running also were listed: a game; the BASIC editor; Amy's job; OCCUPY on Mr. Moran's terminal. I was listed as DCL—digital command language—meaning I was just sitting at the $ prompt, doing nothing. SYSTAT could tell me a few things, like who was where and what they were doing—information we all needed to know, since the computer was shared among us. If a program was slowing down the system, like a game of DUNGEO, a SYSTAT would pinpoint the perpetrator. Heckling might ensue—"Log out!" "You're slowing the system!" But SYSTAT couldn't tell me what I really wanted to know—had BIGBRO logged my aborted attempt at cracking Mr. Moran's account?

I got my bookbag and walked out the door. I pictured Mr. Moran coming into the room, unlocking his terminal, and seeing a flash message on his screen, a bit of automatic mail from BIGBRO. Paranoid mental images coalesced into a fait accompli. Re: Attempted Login. From: BIGBRO. "KB:25 attempted login to 1,0 at 2:35 pm" followed by the date. "Read log-file for details." Then Mr. Moran, pulling up the log of all keystrokes for every terminal for every day of the entire year, would scroll to today and see that I'd been in the room, and he would know it was me because I'd logged in at another terminal moments after the aborted break-in attempt. I would lose my account. Certainly be suspended from the computer room. Maybe expelled.

I had no idea what BIGBRO did. None of us did. BIGBRO was the third line of defense, part psychological, part real. There was BIGBRO on the job list, but what was it doing? Maybe it calculated and recalculated pi to as many decimal places as it could. Maybe it generated random numbers. Maybe it did nothing at all and looped emptily to nowhere, just to create some measure of CPU time on the SYSTAT. Mr. Moran had put it there to intimidate us, to remind us that someone—something—was watching.

I went back to the computer room fifteen minutes later. Mr. Moran was working at his desk. I said hello. He said hello. I sat down and logged in. He kept doing whatever it was he was doing, and I felt a wave of relief. *He doesn't know,* I thought. I kept secretly looking up at Mr. Moran for a while, guilty, but he paid no attention. Although I was never certain that BIGBRO hadn't seen me, I never again tried to break into Mr. Moran's account.

Hacking accounts was bad, but it had been done before my time by students whose names had become legend. One of these was Joel Westheimer, who'd graduated in 1981, before I'd arrived at Horace Mann. I knew his name from the Help files on the system. He'd written a lot of the programs we used: USERS, which listed the names of everyone logged in at a given moment; REMIND, which at a preset time and date would print a little message that you'd entered; and OCCUPY, which locked a terminal for five minutes, enough time to go to the bathroom and get back without another kid taking your terminal. Westheimer's name floated about, mentioned from time to time by the older kids who remembered him, and by Mr. Moran, who thought him one of the best programmers ever to go through the room. It was Westheimer, however, who'd caused Mr. Moran great embarrassment by breaking into a computer owned by Grumman Aerospace.

In 1979 Horace Mann bought computer time on a big mainframe on Long Island owned by Grumman, an aerospace and military contractor. Students used the Grumman computer, dialing in by modem from the old computer club room. One evening, Westheimer and some of the older kids discovered they could gain access to the system administrator's account; they had discovered an "erased" list of passwords when a badly written BASIC program, written by one of the students, crashed. It was a crude, early-version data recovery, the "unerasing" of supposedly "deleted" information. Staring at the password file, Westheimer recovered the password to the Privileged Account and logged in as the System Administrator. This was a major score, giving Westheimer complete control over the Grumman system, access to every account, including accounts from other schools and Grumman employees. So he built himself a triumphal arch: he changed the "message of the day." The MOTD was a little announcement that began "Good morning, Grumman users . . ." and offered general information, such as when the system might be brought down for routine maintenance.

The next day, as hundreds of Grumman users logged in, they were greeted by Westheimer's message.

Good morning, warmongers!

Grumman soon uncovered the source of this affront and Horace Mann was kicked off the system. Things weren't much better when Horace Mann contracted with our local rival, Riverdale High School, for computer time. Riverdale had a PDP before Horace Mann, and once accounts were switched over to the Riverdale system the temptation to break in became even stronger. Grumman was a faceless company in Long Island, but Riverdale was our arch rival: in athletics the Riverdale–Horace Mann games were always the biggest of the year; in admissions, Horace Mann gloated over its persistently higher SAT average and better college placement. But in computers, Horace Mann was grossly deficient. We had to contract with Riverdale to use their computer!

Within months Horace Mann students had broken into Riverdale's system, stolen all the Riverdale student passwords, and left messages in their accounts. This time Mr. Moran took action by banning one eighth-grader, a high-energy, brilliant kid, from the computer room for the rest of the year—a young Paul Haahr. He lost his computer account and could not enter the room because Haahr, like Westheimer, had discovered a way to "unerase" passwords. During his suspension Haahr would sit outside the room and ask people as they went in and out what they were doing. What programs were they working on? What new things had they discovered? Softened by this dedication, Mr. Moran took a liking to Haahr. He let him come back in.

By the time I knew Haahr our desire to crack systems had been cooled under Mr. Moran's careful management. With Super Users overlooking the younger kids, peer pressure blunted temptation. Pranks became more subtle and bizarre, inflicted on teachers outside the room. When Haahr found out I had Mr. Donadio for history class he recounted a prank Westheimer had concocted. It had been Westheimer's idea, and Haahr, who idolized and looked up to him, had been eager to follow.

Mr. Donadio, my ninth-grade world history teacher, had been Joel Westheimer's history teacher. Mr. Donadio was a short man with thick glasses, dark hair, and a habit of sticking his hands deep down the backside of his pants as he lectured. He was a brilliant lecturer. I loved his history class and would sit at the edge of the seat taking notes so fast that my hands would cramp. When he taught, history seemed action-packed, tall tales that were nonetheless true. Mr. Donadio, however, had two peculiarities that figured closely in Westheimer's plan. Obsessed with the stock market, Mr. Donadio would list his portfolio of stocks every day on the board. He also had a telephone in his classroom, which was unheard of at the time. Whenever the phone would ring during class (which it did every now and then) Mr. Donadio would pick it up and say, "I can't talk right now" and hang up.

We figured it was his stockbroker calling.

One morning, when Mr. Donadio was out of the classroom, Westheimer and another student snuck in and opened up the telephone with a screwdriver. They disconnected the part that told the telephone that the receiver was off the hook, and carefully put the instrument back together. Back in the computer room they found Haahr, who would do anything Westheimer asked. Westheimer told him to go into the system room during the next period and dial the number he gave him using one of the incoming modemline telephones. Haahr was told to set the phone on automatic redial. Haahr agreed.

Sitting in Mr. Donadio's class, Westheimer and his friend could barely contain themselves. Mr. Donadio paced back and forth, lecturing with his hands down the back of his pants, every now and then pointing a finger to emphasize a fact. When was the phone going to ring? Westheimer wondered. Maybe Paul got cold feet. Or maybe he was just waiting for the class to go on a bit more. . . . The phone rang. The pranksters looked at each other. Mr. Donadio picked up the receiver before the second ring. "I can't talk right now" he said, and hung up.

The phone rang again.

Mr. Donadio stopped, picked up the phone, and said "hello." No answer. He hung up. The phone rang again. Mr. Donadio shoved the receiver back in the cradle. A few giggles filled the room. The phone rang again. Mr. Donadio picked up the receiver and hung up, again and again. Now the laughter came hard, spreading from kid to kid until the entire class was on the verge of chaos. He turned down the volume on the phone as far as it could go, and now the phone rang, softly, over and over. Mr. Donadio couldn't unplug it—it was an old model, hardwired to the wall; so he dismissed the class. The pranksters ran downstairs, laughing the whole way, to the computer room, expecting to find Haahr. Instead there was Mr. Moran. He'd found the phone off the hook and was wondering what was going on. Thinking Mr. Moran would appreciate the technique involved in such an elaborate practical joke, they told him. Mr. Moran didn't think it was funny. He told them to go and apologize to Mr. Donadio. They went upstairs, unsure of what would happen next—would Mr. Donadio decide to punish them?—and confessed. Mr. Donadio listened. He smiled. To Westheimer's surprise, Mr. Donadio understood that behind the prank was something else: technical sophistication, intelligence. Mr. Donadio cautioned them to use their skills in a better way, and they were forgiven.

Pranks were a part of that room's culture. One fall day in ninth grade I was minding my own business, playing DUNGEO, dealing with the nasty thief who had taken my treasure and hidden somewhere in the dungeon, when suddenly my screen scrolled! Text appeared where I should be typing:

```
H-O-P
>hop
Wheeeeee!
```

Huh? I had just hopped in the air and the computer said "Wheeeeee!" What was going on? My terminal had gone berserk on me.

```
>hop
Very good. Now you can go to the second
grade.
>hop
Are you enjoying yourself?
>hop
Do you expect me to applaud?
>hop
Have you tried hopping around the
dungeon, too?
>hop
Are you enjoying yourself?
>hop
Very good. Now you can go to the second
grade.
>hop
Wheeeeee!
>hop
Wheeeeee!
>hop
Have you tried hopping around the
dungeon, too?
>hop
Wheeeeee!
>
```

I frantically hit Control-C to get out of the game. I hit Control-C again and again until I saw the calming $ prompt, meaning that I was finally out of the game. Then I heard a voice laughing from inside the system room. Amy was looking at me through the glass, laughing so hard at my face, which I realize must have looked horrified. In keeping with the room's ethic, Amy came out and told me how she did it.

That's when I first learned of FORCE, and the human lab-rat experiments you could play on unsuspecting victims—like me. Amy had typed FORCE hop,hop,hop,hop,hop,hop,hop, hop,hop,hop and I had hopped all right, hopped halfway across the dungeon and out of my game. For one weird, sick moment, I had wondered if the computer had come alive.

8
Breaking In

IN THE SUMMER OF 1983, between ninth and tenth grade, my stepfather offered me a job in his advertising agency. I was fifteen years old, too old to return to summer camp in Vermont, and my family felt it was time for me to experience working in the real world. My stepdad, Edwin Lefkowith, was an entrepreneur who'd built an advertising agency specializing in corporate identity design, annual reports, company logos, and packaging for clients that included Chase Manhattan Bank, IBM, and Nabisco. The thirty-five employees at Lefkowith, Inc. worked in sun-filled offices on the top floor of an office tower on Third Avenue and 51st street. Ed gave me a small desk in the windowless office kitchen that doubled as the computer room. I couldn't have been happier. A few feet away stood a six-foot-tall minicomputer, the Microdata, built by McDonnell-Douglas. A system room of my own!

Like the monolith in *2001*, the Microdata, encased in opaque, smoky-brown glass, stood as a silent sentinel, a mysterious object. It was connected to a dozen terminals throughout the agency. I soon discovered, however, that most of the employees avoided the machine. Unbeknown to me, I'd arrived in the midst of a profound transition. In companies across America computers that formerly had been relegated to the back office, doing accounts receivable and inventory tracking, now were moved to the front office. Lefkowith Inc. was no exception. The Microdata had been acquired in 1976 to track client expenses and to run the company payroll. By 1983 the machine also was used to write letters and conduct "personalized" mass mailings. A minority of employees—the technophiles—embraced the machine, writing their own letters and pulling up electronic files on clients at their desks. The rest—the technophobes—avoided the computer altogether. Perversely, in the short-term those least willing to learn were rewarded most: they relied on secretaries to write letters for them and look up client files. Conversely, the technophiles, needing secretaries less, had more work to do. Their reward would come three years later as computer software, modems, and laser printers came to replace the traditional creative process of designing ads on paper. Those unable to use computers would have two choices: quit or learn.

Ed assigned two employees to supervise me. David was the official system administrator. In his late twenties, with long hair and glasses, he was part of an earlier generation of hippie hackers from the days before personal computers. Gail was a part-time consultant who wrote software for the company, customizing applications for their needs. Together, David and Gail gave me a small portion of the Microdata in which I could explore and program. Most of the time, however, I did data entry. Drudge work. Once in a while David would sit down with me and explain how to conduct rudimentary statistical analysis and calculations. Using mathematical formulas he gave me, I'd write a short program in BASIC, taking in data and producing results suitable for printout on the line printer, which used green-and-white perforated paper like the one at school. At the end of the week Ed would sit down with the printouts and go over the company's books. He'd invite me in to watch, explaining how expenses and revenues balanced.

The technophobes saw me as a modern-day whiz kid. Women especially were comfortable asking me questions, and one of them, Barbara, I had a crush on. I'd go to her office whenever I could and help her with any computer problems she had. For the most part, however, my experience with computers in the real world confused me. I found myself frustrated with the Microdata. Instead of an engine of creativity and fun, it was an engine of routine mechanical processing. Rather than exploring, I traveled around a well-marked track. Unlike the PDP or my Atari this machine had no context. There were no Haahrs to tell me stories, no Westheimers to learn from, let alone a Boz to simply sit down and program with. Apart from David, Gail, and a few of the technophiles, most of the employees reacted to computers with disdain, fear, and bewilderment. When the summer ended I was relieved to find myself in the school's computer room, back with a machine and a community to which I belonged, where computers were all right. It was with some surprise then, on the first day of class when Mr. Moran began to address the very thing I'd experienced over the summer—computers in the real world.

"People are going to rely on you," Mr. Moran told us at the beginning of Computer II. "As programmers, you must be responsible for your programs and how they're used." He stood next to the blackboard facing us as we sat in two long rows along the main table. Scott and Aaron had dropped out of computers, and I sat beside Boz. Nearby were Ian and Misha. We'd all been looking forward to this day.

"Most people are in awe of computers, as if computers know the answers to our problems." I stared at Mr. Moran.

He had said that today we'd learn about the history of computers. This sounded more like a lecture on morals. "They think 'if a computer says so, it must be true,'" he went on. *When do we get to log in?* I thought, doodling on my notepad.

"Computers can only give information as good as the instructions we give them." Mr. Moran removed the metal wand from his shirt pocket, extended it, and put a chalk in its tip. He wrote on the board, smoothly: G I G O. Garbage In, Garbage Out. We all knew the word. Mr. Moran turned to face us. "Your job as programmers is to write programs that people can understand. People are going to forget that a programmer wrote the software. They are going to blame the computer if things go wrong, and praise it if things go right. You cannot forget about them. Always think about the user."

That's what we called people—users. Users are the dweebs who use the software you write. Users are not people, they are monkeys, subpeople tapping at the keyboard. Programmers are people.

"Some people say we are at the beginning of the Computer Age. And they think that computers will change society more radically than the Industrial Revolution did. I don't know if that's true." He paused and looked at us. I stopped doodling.

"What that change is going to be will have a lot to do with you. You are the first children to grow up with computers. What you do with them will be up to you. Mastering the tools will not be enough. Being a great programmer will not be enough. Knowing how computers work will not be enough. You will have to know why you are programming, and whom you are programming for. This machine," Mr. Moran pointed toward the PDP in the system room, "exists in the real world. When you leave here, and you will one day, your programs are going to exist in the real world. Being in this class means you are going to be prepared for the real world. And that means being responsible for what you create." He was teaching us about the good side, and the dark side, of hacking. About learning to use our power wisely. Computer II was more than learning how to code, it was learning about the ethics of coding, of the responsibilities that came with what we would learn.

By teaching us how to think about programming as craftsmen, Mr. Moran was imparting to us certain knowledge that we could take our skills out into the world and use them. He would transform us—from power-users who knew how to program and run software—into people who used computers as a way of thinking, to augment our intellect. Programming can be a parrot's art, learned by imitation. My code came from magazines, games from whose pages I retyped into my Atari. But in Computer II and later in Advanced Placement Computer Science, which I took the following year, I discovered programming as a model for looking at other questions, even those outside the realm of computer programming. Problems of all kinds became systems that could be decoded, reassembled, and in the process, understood. What we learned here could be applied before even going to college; although we could do adult work, we were still teenagers, morally unpredictable.

Mr. Moran cautioned us to be wary of temptation, the feeling of power that came with insights we would soon experience under his teaching. The lure of programming is partly the experience of enormous power, of discovering things that might otherwise remain mysterious, hidden. Taking such knowledge out of the confines of the classroom and transplanting it out in the world, with its myriad computer systems, is a potentially flammable act. Programs that affected the real world were a source of mystery and fascination; few of us had seen real commercial software in action, used by real people to accomplish real goals.

That year Mr. Moran taught us the history of computers—how they were invented from ENIAC in 1945 to the present. And we learned what made all computers the same when you were deep inside the system. He showed us the sensitive place where hardware meets software, a place where logic gates control the flow of electrons over circuits and software becomes matter, and matter becomes software. The inside of a computer is like the sun, where light is made. Light is both a wave and a particle at the same time, and the heart of a computer is also a place where everything is two things at once, both ephemeral evanescent pulses of electrons and tangible silicon pathways. We learned about Artificial Intelligence, the ethics of hacking and programming, and the psychology of people and computers. Between the history and philosophy, Mr. Moran taught advanced BASIC, LISP, Ada, EDT, RNO, ASSEMBLY, APL, FORTRAN, LOGO, graphics using the plotter, and the physics that made computers work. In Computer II many of us shared a similar fantasy: to model and map the real world. Once I'd longed for faster, more immersive graphics and fantasy worlds; in Computer II I started fantasizing about my history program, what I called World 2.0.

It bothered me that the way we were taught history—memorizing facts, following the biographies of great leaders, studying economic forces—only gave us part of the overall picture, because it was static. In American history class we learned of Indian Wars, population movements, governors signing declarations, each discrete fragment a link in a chain called time. But the image was illusory. Time never came alive in history class; it was difficult to see patterns or shapes in the facts we learned. Patterns reveal hidden forces; if the pattern isn't visible, the forces remain hidden. If you're too close, events seems chaotic—there's no pattern. If you're too far away, any pattern you see might be misleadingly simple. Yet what if an algorithm existed that modeled the world as a complex system of inputs and outputs? What if you could simulate a real-time animation of changes? World 2.0 let you do that. You could observe the flow of people migrating from one place to another, cross-link these migrations with changing weather patterns, demographic information, political data—there was no limit to how many variables you could input. The question is how to link them, and by linking them, what answer will they produce? This was a different way to think about history.

Writing World 2.0 was impossible; my skills too crude. And even if I were a master programmer, I'd have to somehow get all the data representing this or that change and enter it into the computer, which alone would require a lifetime. But in my fantasies no such limitations existed. In World 2.0 I saw a spinning Earth against the dark background of space. Anything you wanted to learn could be accessed by interacting with the globe. Spinning it counterclockwise reversed time, rewinding history. Spinning it clockwise moved us into the present. Faster still, and the globe would illustrate the future, based on extrapolations from the present. The scale could shift too—from the continental to the local, merely by flying closer to the surface. Then lines and color and words would appear, a thickly textured dance of events with granularity, moving across the face of the Earth. By closing in, detail would increase rather than break apart, as it does when you peer closely at a newspaper photograph or the ASCII art taped above the blackboard. The experience of floating in the heavens looking down felt Godlike. While there existed an element of superhuman power in World 2.0, beneath it are premises that reflect a way of thinking that comes from intense programming, and from being a kid around computers.

My teachers tended to teach history from the perspective of Ideology. History as a narrative of Liberty: over time we are moving toward more democratic governments. History as a narrative of Tyranny: from Babylonia to the twentieth century the economic interests of the rich have required the oppression of the poor. Whatever the narrative—and several could be intertwined at once—their premise is that the world should be looked at from the framework of an existing model. Ideology was the variable that controlled the way we were supposed to learn about our past. IF Ideology = Religion THEN the past means this. IF Ideology = Communism THEN the past means that. And so on. Thinking in ideological terms was second nature at a time when President Reagan could sincerely say that the Soviet Union was an Evil Empire.

Exposure to the PDP, to programming, and to interconnected systems of computing machines made it seem like my other teachers were going about it backward when it came to learning. Instead of looking for patterns to find meaning, they knew the meaning and created patterns to support it. In a sense, I had created an ideology of my own—Systems Analysis—and yoked it to the principles of the scientific method as a model for looking at the world. In the computer room all of us—Boz, Haahr, Hilal, and even Mr. Moran—would have raging arguments about politics, but rarely from dogmatic positions. None of us identified with Democrats or Republicans. Some teachers mistook this for apathy. We did care about solutions to problems in our society, but our approach was different. I imagined that at some future point we would settle unresolved societal issues by modeling problems and looking for different outcomes.

Behind our eagerness to learn lurked a secret. The intense programming, the collaboration, the room itself, had become a refuge for many of us. Divorced parents, divided homes, siblings battling drug addiction—all the strains of home disappeared when we were in front of the terminal or in our bedrooms with the door closed and computer on. Misha, an excellent computer programmer, submersed himself in his algorithms in a fierce bid for independence. He became a computer consultant at the age of sixteen, charging tens of thousands of dollars for a single engagement, with clients on Wall Street. We knew his parents went through an acrimonious divorce, but he would never speak of this. As if to protect himself, he became self-supporting. He would go on to Stanford and pay for his education himself.

Although our parents' divorce had not been acrimonious, Samantha and I also felt strains. Early in our joint custody experiment I lived with my dad and saw Samantha only on weekends. One weekend she would stay with me at my dad's, the next weekend I would stay with her at my mom's. Samantha and I had always lived together, and though we'd fought "like cats and dogs" as my mom used to say, behind our battles we loved each other. Living apart from Samantha during the week, I didn't see the inexorable slide she made from being a "good girl" to becoming a "bad girl." Because she went through puberty very early, developing breasts at the age of ten, older boys recognized her budding sexuality and their attention drew her forward. I watched, with amazement and dismay, how she went from first base at eleven to second, third, and home by twelve. I was both concerned and, secretly, a bit jealous. I was still a virgin and my little sister wasn't. I'd been on a similar course at age twelve and pulled back. She didn't, and our paths diverged dramatically. As I got older I became a better student, more popular, more connected with the mainstream at school. As Samantha got older her grades dropped precipitously. By sixth grade she was no longer a good student. The girl who'd won first-prize in her Parisian school for academic excellence was expelled from seventh grade at Hewitt, the all-girl school my parents had her transferred to after Fleming.

Samantha started to change when we came back to America. Instead of fighting her difference, she accepted it and went further into it, embracing it. She wanted to be different. I can mark the turning point—in fifth grade, when she was ten. I was in eighth grade, just discovering my Atari. I had found my place, and she had lost hers. I remember one time she came home from school so sad. My mom had bought her a pink T-shirt at Fiorucci, a disco-fashion emporium that sold sequined outfits and brightly colored clothes. On the front of the T-shirt was a copy of the rate sheet that's stuck to the door of New York taxis. It was a smart T-shirt; it had a kind of witty coolness my mom liked. But at school it turned into something else. The kids began to taunt her with remarks like "How much do you cost? How much for a ride?" and my sister was badly shaken. She had small breasts then, but in fifth grade that was strange, because most girls didn't have breasts at all, and the kids teased her. It's hard to say who changed whom—if boys looking at her made her a woman too soon, or if she became that way because that's the way it had to be. She made a decision then, since no one wanted to treat her like a girl anymore: she'd become a woman.

Samantha started smoking and hanging out after school
with a few other girls who did the same thing, behavior that
coincided with both our parents' remarriage. In 1982 our
mom and stepfather married. In 1984 our dad married our
stepmother. Now Samantha had more than an older brother
to compete with for time—she had two new adults. In this
complex, evolving world of four parents and two children
we all decided to enter into family therapy together. While I
was merely angry and silent, bent on building an extra life
at school, Samantha's anger led her outside of school, to
building a life in the streets. She and a few girls became the
bad ones, and it all happened so fast. She lost something
along the way, much too early. Her childhood. A faint memory of a time, she would later tell me, that disappeared
somewhere between age ten and eleven.

*   *   *

In tenth grade I was getting A's in Computer II and my academic prowess had begun to filter outward. My grade average improved from C to B–. In the wooded hills of Van
Cortlandt Park I surprised myself, and coach Beisinger, by
running 2.5-mile races with times close to fifteen minutes;
the best runners had times of fourteen minutes. Several
days a month I'd stay after school to work on the Governing
Council as one of my grade's four representatives. I'd been
elected on a non-ideological platform, the centerpiece of
which was lowering the price of yogurt in the cafeteria. Taking a page from what I'd read about politics—"all politics
are local"—I decided to emphasize quality of life issues like
an improved student annex for studying rather than high-
concept ideas like mandatory teacher evaluation forms. My
inspiration had been Paul Hilal, who at the end of the previous year had been elected Student Body President; Hilal

was also a Super User and a rigorous programmer. Paul Haahr and I had both stayed late after school helping Hilal silkscreen his campaign T-shirts. Where Haahr was eccentric and long-haired, Hilal was clean-cut and outgoing. That fall, in 1984, I remember watching Hilal as he spoke to the General Council. Hilal's election symbolized how rapidly the social isolation of the computer room was changing. In the seventh and eighth grades a trimester of computer study was mandatory, and science classes at that level required some skill in BASIC programming.

In my first trimester of tenth grade the cafeteria found a new yogurt distributor and the price fell to 90¢ from $1.10. My constituents were happy. So was I. Like a virus, my enthusiasm and success with computers had spread. I developed an addiction—computer programming, gaming—that had positive side effects. Samantha developed an addiction too—to drugs, eventually—that would spread, eroding and destroying rather than bolstering and creating. The stress in our family increased as Samantha pushed the boundaries further and further. I reacted by separating myself emotionally, and there was no more welcoming and comforting place to go than the computer room.

Hanging with Misha and Boz and Haahr, what truly mattered was how well you programmed. There was a kind of peace in programming, a place to be at school that was what I imagined religious people feel when they study scripture. It was work and it was pleasure and it could be everything, taking you away to another place. Mr. Moran had a way for us to go, and we started by going to the inside, like medical students, sticking our hands right onto the heart of the machine and feeling it the way we'd never felt it before.

One day in November we were in class, settling in, when Mr. Moran told us to log off. "We are not going to be using the PDP," he said. He got up from his desk and walked over to the IMSAI 8080, a big metal box we called the dead computer, the same machine the kid in *War Games* had used. It looked like an electric guitar amplifier, colored blue, the kind of wacked-out piece of gear that Scott's dad would've built for rock musicians a generation earlier. Frankenstein machinery that scares most people because it isn't user friendly. I thought its rawness was why it was excellent. No one ever used the IMSAI—a few times I'd seen Haahr playing with it—but most people didn't go near that machine.

Mr. Moran turned on the IMSAI, flipping a red switch. A row of small red diodes, *Star Trek* style, flashed on the front panel and then dimmed. This computer, Mr. Moran said, is just like the PDP, which is just like the Apple II, which is just like an IBM mainframe. Now this one might be smaller, he told us, pointing to the IMSAI, and a lot less powerful, but inside the architecture is the same as any other digital computer. Because the IMSAI is designed to work without a monitor or a keyboard—you can program it by flipping the little switches along the front—it lets you get closer to the microprocessor. Close, close to the machine, brain to brain. Mr. Moran showed us the way to load bits using the switches. Each switch corresponded—was *hardwired,* he said—to a specific point in memory. By flipping the switch the bit is either "on" or "off," 0 or 1. This was real, nose-in-the-bits programming. Without all the layers of "interfaces" (such as menus and characters), but instead just switches hooked right into the IMSAI's central processor, the distance between my brain and the IMSAI's 4K brain was miniscule. I found the closeness exhilarating. And that week I figured out a really neat hack.

I discovered that you could program the IMSAI to make the diodes under each switch flash on or off in patterns by making programming loops to do it. Depending on how I wrote the loop, the lights could pulse all at once, or one by one, or with a delay, creating the illusion of a streak or wave. I wrote a short program to flash the lights in patterns. I learned by fiddling interactively with the machine. I began simply, turning one light on and off, then on and off one by one, and finally, by holding some switches on and some off, I started to make beautiful lines. It was a strange hack few could appreciate. To an outsider, like my mom, lights going on and off on a metal panel is not as exciting as, say, an elaborate animation on my color Atari. But on the IMSAI, which was not designed to display graphics, using the diodes to do this has a certain elegance, a kind of useless usefulness. That's the key—it's not useful. It's self-expressive, a quiet thing done between you and the machine, simply because you can do it and because it's fun. I could tell Mr. Moran was happy when I showed him my diode hack. He smiled. The pleasure in this reflected one of our Commandments: computers are tools of beauty, makers of art.

Paul Haahr taught me how to play Ping-Pong with the switches. He explained how you could change the program to detect whether the last switch on the left is up or down, and whether the last switch on the right is up or down. You could program the diodes, based on the position of the switches, to light up in series, forming a moving "ball" that bounced between the two sides. Of course the game wasn't difficult, since the ball went down a single line. But as Paul and I increased the speed of the flashing lights, timing the switch was hard—not too hard, but hard enough that the game got real for a moment. And that was great. Suddenly we were hyped up, playing a game on a crazy old piece of gear. Our game reflected another Commandment: systems, however old or new, are to be explored, and all information shared.

This was around the time that the lures of dark-side hacking—of breaking into real-world computers and telephone systems—became most difficult to resist. Now that my skills were increasing I had something to offer other kids, who in turn might have information to share with me. Throughout the city there existed an informal network of like-minded teenagers, all of whom found thrills in sneaking where they didn't belong. That summer, while working at Lefkowith, Inc. I'd met Joey Ferraro, a hacker from Bronx Science, at a party in a big, rambling West Side apartment, with books everywhere and large hallways and a huge kitchen. I'd been in the kitchen speaking to my friends and some kids I'd just met when Joey walked in.

He had long shaggy black hair and a Van Halen T-shirt. Strangely, he was carrying a bookbag slung over his shoulder, as if he'd just come from school. I asked him if he was in summer school; was that why he was carrying textbooks? Casually, he dropped the bag on the table, opened it, and pulled out a lineman's telephone: a big orange handset, with the touchpads in the center, and alligator clips! He then pulled out knotted wads of wire and plastic boxes and other electronics I didn't recognize. Between us were a few beers, and we started exchanging information.

Joey knew about telephones . . . he was a phone hacker; his system was the vast network of telephone switches that connects the world. He laid out a list of BBS numbers. There, he told me, I would find files—or *phi*les, as in *ph*ones—with information on getting free calls: MCI calling card generators to get long distance, 800 numbers for PBXs with outside line access. Joey turned around and unscrewed the phone. He popped it apart and placed his alligator clips on the wires. He was testing the line. The rest of the kids in the room looked over, some of them too drunk to understand, and suddenly

... BOOM! Wisps of smoke came out of the phone! "Shit!" I said. The others howled, joyous bedlam. I didn't know who lived here, and I was suddenly afraid. Joey didn't know who it was either and didn't wait to find out. He shoved his belongings back in his bag and with one swoop of the arm threw the bag over his shoulder, opened the back door in the kitchen, and headed down the back stairs. Caught in the moment, I jumped up and impulsively followed Joey, slamming the door behind me, and ran and ran down ten flights of battleship-gray stairwell, dizzy and a bit drunk when we hit the night air outside on a side street a half block from Central Park West. We were laughing, retelling the moment through gasps: Imagine coming into your kitchen and seeing your phucking phone on phire. Cool!

We hung out by the stone wall of Central Park drinking beer. Then Joey and I made a deal. I knew about PDPs—our school computer—and most kids didn't know anything about how our operating system RSTS/E worked or how to set up Super User accounts to get a privileged bit (the technical name for access to other people's accounts). PDPs were popular in the 1980s with universities and companies. If you knew PDPs in the loose-knit hierarchy of high school hacking, you had valuable knowledge. The first rule of hacking was hands-on access, and few among us could claim the sort of access I'd had to PDPs. I would give him PDP info, he would give me phone numbers. A fair trade.

As Computer II got under way in October I modemed Joey files I thought he'd find interesting. Nothing too sensitive. Mostly Help files and public programs. Joey would upload the stuff I gave him onto local BBSes run by self-styled kid hackers from their bedrooms, getting credit, looking good. Most kids didn't know much—certainly much less than Joey—and they found the information I gave him exciting.

Most of his audience just had a few BBS numbers, a dumb handle—Mazter Craker, Darkphile—and the hacker files they read were mostly junk. Joey, though, was good. One day he asked if I'd let him use my account at school.

"Can I call in? I won't do anything bad, I just want to look around."

I wasn't sure. Certainly if Mr. Moran found out I'd get in deep trouble. I laughed out loud. "Yeah. Whatever."

Joey persisted. He offered me something in return—COSMOS. COSMOS was a hot topic of discussion that year. People thought it was a code name for a computer that controlled the telephone system. I challenged him—COSMOS doesn't exist—and Joey took out a pencil and wrote down a phone number. "Dial this," he said, "and call me later tonight."

That night I booted up my Atari, loaded my Chameleon terminal emulator, set my modem to 8 bits, no parity, Xon–Xoff, and a minute later found myself staring at what was unquestionably a New York Telephone login screen. It asked for login name. I didn't have one, and the moment I tried a random guess by typing "help" (which on some systems triggers the Help system) I was disconnected. Good security. I thought about it—a real-world computer. A telephone company computer. I wanted to go inside. I could give Joey access to my account. After all, there wasn't much danger to it. LILBRO and BIGBRO were watching. I didn't have a privileged bit. Joey would be penned in. I called Joey and gave him my password.

Fifteen minutes later I was very disappointed. The computer was filled with complicated telephone company Help files. It was a library of some kind, a place for technicians to look up basic information. An unfair trade. Our PDP was far more interesting, filled with games and interesting programs. I logged off and dialed Joey's number.

His line was busy. I called Joey every half hour, for hours. Busy. I became anxious. Near midnight, Joey called me.

"I think I crashed the system," he said.

Oh.

"I'll call you back," I said.

I dialed Horace Mann. It rang. No answer. I tried again. No answer. I called Joey to ask him what he'd done to crash the system and he said he didn't know. I thought about BIG-BRO and LILBRO watching the dialups, seeing me log in—or rather Joey logging in as me—and I thought tomorrow would be a good day to be sick.

The next morning I felt like throwing up the whole way up to school. As I ran up the hill nothing else mattered other than getting to the room and logging in. I arrived at the computer room door and looked through the window. The room was empty. I twisted my neck to look into the system room. Mr. Moran was in there. So was Paul Hilal. With his new duties, Hilal didn't come into the room as often. Knowing he was in there with Mr. Moran scared me. Things did not look good, I thought. They were looking at printouts together, and they both seemed upset.

I crept in and sat down at the first terminal. Blank. They were all blank. The shades were still down; the room was gloomy. I flipped on the fluorescents. Mr. Moran and Hilal looked up from the system room, seeing me. Hilal came out.

"You can't log in. The system is down," he said.

"Why?" I asked.

"It crashed around six o'clock yesterday evening, and no one knows why."

"Were any files lost?" I asked, praying he'd say no.

"We won't know until we get it back up again. We're booting off tape, and we'll check the backups. You want to help?"

"Uh-huh."

"Just sit out here, and tell the kids coming in to leave."

Hilal went back into the system room, and I dreaded the moment when they got the log tapes back up. Mr. Moran would see I was in the system at the time it crashed; at that instant, I thought, "If I was logged in last night at six, when the system went down, *of course I would know it was down!*"

Boz came in. He seemed too happy, grinning like he knew something. "What's going on?" He looked at Hilal and Mr. Moran in the system room. Boz liked intrigue.

"System's down."

"Oh? What happened?"

"I don't know."

I figured I'd never be Super User now. Boz would. Suddenly the terminals flickered on.

"Cool." Boz logged in.

I logged in.

"Password incorrect," it said.

What?

Mr. Moran and Hilal came out of the system room.

"What happened?" Boz asked.

Hilal explained. "It looks like a job last night somehow ran the system out of memory, and instead of kicking the job out, the system crashed." Mr. Moran would file a bug report with Digital. We were a beta site for RSTS/E, and the current version of the system was in "beta," or pre-release testing. We were expected to report strange incidents to Digital as part of our beta testing agreement with them.

Feeling confident, I asked "Do you know what job did it?"

"No. We can't tell," Hilal replied.

"Was anyone logged in?" Boz asked.

"No," Mr. Moran answered. "No one was logged in."

No one was logged in? Joey was logged in. I was shocked. *He is good,* I thought. Covered his tracks. How did he do that?

"Mr. Moran," I said, "I changed my password and forgot it. I can't log in."

Laaaaame. Boz laughed at me. I couldn't admit that someone else—Joey—had been in my account and changed it. "Sorry," I said sheepishly. Mr. Moran logged into his terminal at the head of the table and asked me to choose a new password. I then went to another terminal, further away from Boz, and logged in, dreading what I might find in my account. I hoped Joey hadn't left Godzilla-sized clues all over my directory.

Nothing.

Nothing but my programs.

But wait. There was one. I saw one file. A text file. MORDOR.TXT. Joey liked the *Lord of the Rings*. Mordor. The evil empire.

"Where shadows lie," said the first line.

And then a bug report! Joey explained what he did to the system, rubbing my nose in it. *Confident fuck,* I thought. He must have written it before he crashed everything. I just scanned the lines, quickly. Digital sure would like to see this.

I could give it to Mr. Moran.

I could delete it.

DEL MORDOR.TXT.

File deleted.

Joey and I stopped talking after that. In the end he didn't do anything bad: nothing was lost, and no one got in trouble. But the scare with Joey made me think about being serious in the computer room. It made me realize that if I wanted to be Super User, I still had a lot to learn.

9
Symbiosis

FEAR OF PUNISHMENT prevented me from revealing Joey's hack to Mr. Moran. Joey, however, had acted fairly, within the parameters of our code. He hadn't merely crashed the PDP: he'd explored the system, found a vulnerability, and reported it back to me. My deleting his bug report prevented the rest of the room—perhaps even Digital Equipment Corporation itself—from understanding an imperfection in the PDP's system. Elsewhere in the world, on another PDP, another hacker might discover the same vulnerability and unlike Joey, exploit it maliciously. What responsibility would I then have for not forwarding the information to Mr. Moran? Information that, whatever my punishment for letting another person use my account, might have protected another system from attack? These prickly ethical questions were best avoided by acting in good faith and not breaking the computer room's rules, such as the prohibition on sharing computer accounts. Nonetheless, questions of ethics permeated the computer room, because, all else aside, we shared the same computer.

By late 1983, when the incident with Joey took place, several hundred Horace Mann students had computer accounts. The majority of these users were younger students—seventh- and eighth-graders—who were more likely than older students to own home computers. These kids had been in fourth or fifth grade when they first acquired personal computers, and by the time they were old enough to take Computer I in ninth grade they'd spent four years using them. But the older kids in the room knew that the quality of programmers was decreasing, year by year. The best programmers in my grade—me, Boz, Misha, and Ian—were not as good as Haahr and Hilal; in turn, Westheimer, Bruckman, and their confederates possessed skills few of us had.

In part such a decrease in quality programmers reflected the growing sophistication of software, which had become slickly packaged as a finished product as more and more people used computers. The era of frenetic experimentation, of relying on people's skills and creativity to compensate for weaknesses in the software, was ending. The first generation of home computers required some ability to program, but succeeding generations did not. While improved software and technology broadened the computer's appeal to millions, it also made hands-on experimentation by kids more difficult. Instead of rewarding an understanding of the way computers worked, the new machines rewarded the ability to use software. It was the beginning of a move from producing to consuming, from making programs to using programs. In the world outside the computer room, the evolution of human-computer relations had taken a turn in a new direction. More and more that room on the third floor of Tillinghast seemed an oasis, fertile ground in a widening desert. Representing the changes were IBM and Charlie Chaplin's little tramp.

When IBM launched the IBM-PC in the summer of 1981, they'd sold the machine as a computer for the rest of us, one so easy to use that even the little tramp—icon of common sense—found himself charmed by the machine. Leaning on his cane, wide-eyed with joy, the little tramp had found a soul mate in the cute IBM-PC, which sat on a round table beside a rose in a little vase. The ad campaign's message was that with an adult in the field—IBM—it was now all right for grown-ups to buy home computers for themselves. Mature, sensible, and user friendly, the IBM-PC came from a company whose initals were synonymous with commitment, long-term investment, stability. The advent of the IBM-PC—which cost between $3,000 and $6,000 at the time—meant that home computers had grown up and entered the real world. Led by IBM, the market grew quickly. In 1982 IBM sold 128,000 personal computers; in 1983 sales rose to 800,000, and IBM produced more personal computers than any other company, capturing 28 percent of the market. Apple, the runner-up, shipped 600,000 computers, giving them a 17-percent share, down from 29 percent in 1981. With IBM's success came "clone makers"—led by Texas Instruments and a new company called Compaq Computer. By the time I was in tenth grade, the IBM-PC had utterly changed the world of home computers, homogenizing them and reinventing their image as tools for work, rather than whimsy, fun, or exploration.

IBM's dominance led to a sanitization of the software industry. Where once programmers were seen as rebellious, creative types, hacking new programs at home, dreaming of a breakaway hit, the industry now depended on software sold as closed "productivity packages." Kids like the fifteen-year-old executives of Software Innovations who had written games for the TRS–80 had no place here. What company would buy a spreadsheet or word processor from a fifteen-year-old? Learning by doing was harder and harder in a world of "grown-up" personal computers. As the IBM-PC increased in power, programs no longer came in "source code." You couldn't learn how to write a spreadsheet by reading the VisiCalc program. Nor could you discover how a word processor, like Wordstar, functioned through hands-on interaction with the code. Peeking into software was bad—an infringement of copyright, a violation of trade secrets. The new computer user should be a master at using a product—an ace VisiCalc formula-writer—rather than mastering the computer itself. People were expected to accept a layer of mystery between themselves and the machine. A customer shouldn't question why Control-F6 opened a document. A customer shouldn't be able to customize the product, altering keystrokes to initiate a command, let alone program an entirely new function.

Atari withered fastest in this new environment. By late 1983 Atari had fallen precipitously from its position as the world's largest home computer manufacturer. A company that once had stood for innovation, whose name inspired a group of forward-thinking congressmen to call themselves "Atari Democrats," had embraced its antithesis. Their games had been the most creative and trendsetting, but by 1983 Atari emphasized clunky tie-ins, spending $22 million to acquire the rights to *E.T.*, the game, as if by mere association with the film they'd be guaranteed a surefire hit (the game was a flop). Their computers and game consoles, formerly the most advanced, were now woefully out of date. In the first quarter of 1982 Atari had earned $112 million dollars; in the first three quarters of 1983 the company had *lost* a staggering $500 million. Atari's failure came from a strategy rooted in top-down mass marketing. They were trying to change a fundamentally creative tool into a VCR, or cassette player—without even a Record button.

Quelling hands-on access to software choked off the possibility of its continued existence as an evolving, collaboratively produced medium. Open systems like our PDP encouraged a different model of interacting with computers: software was group-designed and improved continuously, and breakthroughs were shared by all. Updated programs were passed by magnetic tape from institution to institution, and each version came with visible source code that invited yet another person to add new functions and more value to the software. In the beginning software was treated like scientific research: an endeavor owned by none, shared by all, to promote the greater good. Uncompiled source code, passed from machine to machine, community to community, found itself subject to the rigors of peer review and experimentation. The results were new and improved programs, which in turn were subject to another cycle of examination and improvement. Because the best programmers tend to be young (people eventually burn out from the intensity of coding), this rigorous process—which necessarily involves the will to learn and desire for change—made its way into colleges and graduate schools. A network of self-sustaining, innovative open systems had thrived for twenty years—from the early 1960s to the early 1980s; now tens of thousands of people were using computers in universities and research labs, with new recruits coming in every year. Ironically, the personal computer—which ultimately created a market force strong enough to suppress the earlier approach—originated from that community of people who believed in hands-on access and open systems.

Nolan Bushnell (who founded Atari), along with Steve Jobs and Steve Wozniak (at Apple) each had encountered computers in university labs. In the late 1960s these three men, who were most responsible for propagating the home computer, spent their early twenties with hands-on access to time-shared computers like the PDP. (Bushnell at the University of Utah, which had a pioneering computer science program in computer graphics; Jobs and Wozniak at Berkeley and Stanford, respectively.) The original urge to create personal computers came from a desire for greater access, more time with these fantastic time-shared machines that were a radical break from the previous generation of computers built by IBM.

The history of computers from 1945 to 1993 fell into three stages with a fourth beginning in 1993; I was in the third. The first, represented by IBM (from approximately 1945 to 1960), was an era when computers were extraordinarily inaccessible, expensive, and hard to use. Programs were fed with punch cards into machines that cost millions of dollars and were operated by a cadre of technicians. Prompted by this new technology, a regime of bureaucrats controlled access to the fantastically expensive machines. IBM dominated the market for computers at that time, and because access was so difficult, improvements in the system came from the company. There were no college kids tinkering, hatching up improvements, because for the most part these machines cost too much to operate to permit "playing."

In 1960 the era of the minicomputer began. This second stage was dominated by Digital Equipment Corporation, which manufactured the line of PDP minicomputers, beginning with the PDP–1 in 1960. Costing hundreds of thousands of dollars, and later tens of thousands of dollars, rather than millions, these machines broadened hands-on access to computers. The PDP would prove so popular that one model, the PDP–11, would be manufactured from 1970 to 1997 (when Digital Equipment shipped the last PDP-11 in 1997 they had $1 million in back orders for the machine). In an industry where conventional wisdom now says technology must be obsolete in approximately 18 months—the time it takes for "Moore's Law" to double the speed of a microprocessor—what can we make of a computer system whose lifetime spanned three decades? The answer rests in the usefulness of systems evolution from the bottom up, rather than incremental product development from the top down.

People bought PDP–11s for twenty-seven years because the machine came with software and a system whose components came not from a single mind, or even a team of programmers from one company, but from the vast collective of thousands of computer rooms around the world. While certainly much of the software on the PDP such as RSTS/E (our operating system at Horace Mann) came from companies who charged for it, much of it did not. Perhaps the finest example is UNIX, which was collaboratively developed over a span of twenty years, beginning at Bell Labs where researchers working on a PDP first hacked out the code. UNIX has since become one of the world's most popular operating systems, and it is a core element behind the explosive growth of computer networks like the Internet. Because it's possible to access the source code for UNIX, it never stops improving. There is no single version of UNIX, but rather myriad "flavors" whose suitability depends on the user's needs. If the users have needs UNIX can't fulfill, they can always create another flavor.

The era of collaborative software production reflected a specific breakthrough in the architecture of computers, what people called "time-sharing" and "interactive computing." Because machines were cheaper more people could use them, so designing methods for sharing the machine's processing power and disk space was essential. Otherwise, if only one person could use the minicomputer at a time, few institutions could justify the expense of owning one.

Time-sharing is the most important innovation in the history of software, because it fundamentally changed how people and computers worked together. It became possible to experience computers, as one early computer scientist noted in 1960, in a "conversation-like mode." In other words, with the advent of time-sharing, people could "interact" with computers for the first time, receiving "feedback." Feedback, interaction, conversation—these are all hallmarks of communication, and communication is, essentially, a continuous process of exchanging models from one speaker to the other.

With time-sharing, and the new concept of operating systems that came with time-sharing, it became possible to work with computers through trial and error, in much the same way people work together. Where once people had to conceive a problem (say, a calculation representing the flow of air over an airplane wing), program the solution, and then run it, now it was possible to begin with a hypothesis, sketch out a solution in code, run it "interactively" on the computer, see what worked and what didn't, alter the code, and run it again until a solution was reached. Rather than furnishing a fully rendered program, the new systems inspired people to begin with an idea of where the solution might reside—a hunch—and explore from there, using the computer as a partner to find a solution. In the environment of interactive computing, making mistakes was good. Starting over was good. In the previous age of batch-processing such activities had been prohibitively expensive, and thus all but impossible to undertake. The new breed of computers thrived on this way of working, and they evolved from souped-up calculators to problem-solving partners.

The third era of computing began around 1977 with the release of the first mass-produced home computer, the

Apple II. Microprocessors had become so cheap that it was now economically feasible for one person to have total access to an entire computer. The economic necessity of time-sharing had ended. For another few years home computer users maintained the open, transparent, hands-on practices that had evolved around time-sharing. Mass produced and sold by the millions, however, home computers were more objects for individual consumption than channels for collaborative innovation. Learning the virtues of group-designed software became much more difficult. Group-designed software marketed itself through word of mouth and firsthand experience rather than magazine advertisements. And since personal computers were isolated, stand-alone machines, people were likely to learn how to use them by themselves (or in the office, as part of work). In the age of the personal computer, software innovation slowed to a process of fine-tuning existing, time-shared operating systems for the home computer environment.

MS-DOS, the operating system on IBM-PCs, came from CP/M, which, in turn, borrowed heavily from time-shared systems like RSTS/E. In 1984 the Apple Macintosh's breakthrough graphical user interface, with icons, windows, and a mouse, came from a fifteen-year-old experimental operating system that had run on a time-shared computer in 1968. By 1984 personal computers had become intimate, everyday devices; they were now part of mass culture. Much of the focus in this third phase was on the development of interfaces that enabled people with little computer experience to learn software quickly. Over time the pitfalls of a top-down, mass-marketing strategy became apparent. Companies like Microsoft became enslaved to their own architecture, MS-DOS, and even with the advent of Windows a certain limit of functionality was reached. Going beyond required too profound a shift from investments, was too risky, and consumers slowly began to grow puzzled. Why was it "impossible" for their PCs to do certain things? Where a group-driven, bottom-up approach like UNIX continued to evolve, top-down PC architectures stalled in the mid-1990s.

The latest phase, which began in 1993 with the release of Mosaic, later known as Netscape, a graphical interface for accessing the Internet developed in a research lab at the University of Illinois, marked a return to the ethos of time-sharing. Personal computers became networked computers. In 1993, as in 1960 and 1977, the terms of human-computer relations took a sudden and profound turn. At stake throughout remained the same question—the form of symbiosis between people and their computers.

Symbiosis was a concept I first learned from my ninth grade biology teacher, who explained that in nature organisms often cooperated across species, mutually improving chances of survival. Humans, he explained, were the only animals to cooperate with their tools—to live symbiotically with machines. Where in nature a bird and rhinoceros help sustain each other—the bird feeding on harmful insects living in the rhino's skin—in society a machine and person can share a similar relationship. Early computer scientists certainly were aware of this and wondered what the terms of human-computer symbiosis would be. Part of the ongoing struggle with computers is defining the terms of our engagement. What does symbiotic existence with a computer mean? Initially it was perceived as the symbiotic relation between master and beast of burden. Computers in the 1950s were seen as slaves and a future was imagined where these colossal servants would pamper and coddle (sometimes in the form of robots, which were ambulating computers). With the beast of burden image came the fear of slave revolt, the Helots rising, as in *Alphaville* and *2001*. In the 1960s with time-sharing a new symbiotic model appeared, one of enlightenment, that was closer in spirit to the idea of the computer as shaman, a guide to knowledge. On the positive side of this image is the idea of journey, of a means to knowledge through exploration; the reverse is the idea of black magic—the computer as a hypnotic, hallucinogenic escape whose spells are understood only by the sorcerers. Initially the symbiotic model of humans journeying with computers to find knowledge was reinforced with the first generation of home computers and in the computer room with the PDP. Later, as personal computers became more advanced and opaque, people began to question whether they promoted wisdom or merely a cacophony of information.

*   *   *

By tenth grade, as my programming skills improved my Atari became less interesting. At home I had no audience, no one to share with or show my programs to. Where the rest of the world purchased home computers at a frenetic pace and obsessed over the latest business tactics emanating from Silicon Valley, at school we became increasingly isolated from outside changes. I paid little heed to how different our computer room had become from those at other schools, where Apple computers led the way into classrooms. Like students at a yeshiva in the midst of the modern world, we continued in our way, hewing to the primacy of shared code, group exploration, and collaboration.

What we experienced had been felt by previous generations, early computer pioneers, the first wave of hands-on hackers. They too had discovered marvelous insights, producing new uses for computers in similar environments.

Ours was merely the latest iteration of a pattern. This particular computer culture was inherently subversive; it was difficult for institutions to contain and promoted alternative structures, communities, and visions that differed from those of traditional hierarchies. This ethos was carried forward through the 1960s and early 1970s by people who reinterpreted it in their own fashion. They did not follow a prescribed game plan. Instead they were inspired by the contagious idea of close collaboration and intimacy with computers as partners in creative thinking. In turn, many believed that symbiosis with computers would decentralize power, devolve the once secret world of planning from central authorities into the hands of the people—which was the unarticulated principle behind our hacking and the way we taught each other how to program.

Yet even with the PDP, the actual need for student-led improvements to the system was diminishing. The system patches written over the years held, and new versions of RSTS/E required less hands-on tinkering to fit our needs. The curriculum had coalesced into a sensible series of classes; the early days of Mr. Moran having to design a pedagogy on his own had faded. That subtly altered the cooperative relationship between students and teacher. Before, Mr. Moran *had* to let kids write system code, otherwise he'd have to do it, which was impossible if he wanted to teach classes as well. In tenth grade Mr. Moran no longer needed Super Users to do fundamental system programming. Nevertheless, our structures remained in place. There was no question of ending the tradition of Super User and its junior, Senior Level User. After all, giving us control had always been partly practical, partly idealistic. The ideal held firm.

In the fall of 1983 Mr. Moran decreed that everyone in Computer II had earned the right to the title of Senior Level User.

As an SLU I received extra disk space and computer memory. That permitted each of us to write longer, more complex programs, essential materials if we were to graduate to the next level of sophistication. The day it happened, Boz, Misha, and I were elated. Now we could begin work on the first big project of our computer careers—what Mr. Moran called the Spy program. The day he gave us the assignment, Mr. Moran stood before us and held up Joel Westheimer's Spy program, saying it was the best one ever written. The printout, nearly four years old, had "WOW!" and "A + + +" written on the first page in Mr. Moran's writing. Staring at the code, I knew that each of us silently pledged to better Westheimer.

Doing so proved difficult. Spy culminated our instruction in BASIC by introducing new concepts ill-suited for this cumbersome programming language. Foremost was the idea of tables—databases containing information on our spies. As spymasters of an unnamed espionage organization, each of us, through our Spy programs, had to manage the deployment of secret agents around the world, from New York to Mexico City, Tokyo, Moscow, or anywhere else a mission called. Need a fluent Russian-Chinese speaker to conduct an assassination in Kuala Lumpur? No problem. Through my program, I could check on which agent fit the needed profile. Each agent was a variable, a "record" in my data table, and its profile existed in "fields" of age, experience, fluency, languages, and so forth. There were also fields describing his current location, the duration of the assignment, and whether assistance was coming from another agent. As the agents conducted missions, fields relating to experience, age, fluency, and so on were altered. Spy, wrapped in a fun narrative, demanded a firm understanding of database management, control structures, and interface design. Mr. Moran insisted that the "spymaster"—meaning himself, as he graded our programs—was a dunce, a fuddy-duddy butterfinger who needed loads of help, on the fly, from the program. At the same time, the technicians maintaining the program were expected, on the inside, at the level of code, to maintain equally neat and well-detailed "comments" explaining what each section did and where it began and ended, so that future programmers could build upon the code without wondering what did what.

Spy was both farewell and completion of our time working with the first programming language any of us had learned. BASIC was what I knew best, the language with which I was most comfortable, but it had its flaws, which prevented us from attaining the next level of fluency. Much as language shapes the way we see the world, certain tongues having no words for certain concepts (and many words for others), likewise programming languages differ in their grammar and vocabulary. One who has learned to see the world through English may have a hard time learning to see it through Japanese. Similarly, a programmer who's learned to use computers through BASIC may find a new language like Pascal difficult to understand because it contains concepts that do not exist in the mother tongue. For instance, Pascal permitted many variants on the familiar IF-THEN statement that was a staple of BASIC programming. Where BASIC might say IF X = 1 THEN Z = Y + 1, Pascal allowed phrases like:

```
while thisistrue do
case terminaltype(0) of
repeat move_to(1,1) until false
```

Each of which were permutations of the traditional IF-THEN statement fine-tuned for particular uses. For the novice, Pascal's variations on IF-THEN could be bewildering.

BASIC's virtue was its immediacy of reward. With little effort one could learn elemental commands. Even the uninitiated could look at a program and understand the code, as in this simple program.

```
10 LET X = 1
20 PRINT X
```

BASIC offered tangible, immediate rewards. It had been designed for teaching beginners how to program (its name stood for Beginners All-purpose Symbolic Instruction Code). Push BASIC further, beyond simple programs of a few hundred lines, and its rewards soon diminished. It was all but impossible to code truly ambitious programs in BASIC. Spy attempted to do the impossible—use BASIC to write a complex program—and by so doing delivered several payoffs. It required all my skills in "flowcharting"—conceiving the control structure of my program—forced me to code in a neat, "modular" fashion, and no matter how well I wrote, revealed the need to go beyond BASIC.

Had I tried to sit down and simply begin coding, my Spy program surely would have failed. Instead I had to begin with a careful sketch of all the various "modules" or "procedures" that would process information and pass it on from one part of my program to the next. When I showed my flow chart to Mr. Moran he asked if I could break down any of the modules into smaller modules. He wanted me to think of the smallest useful unit and build the program up from that scale. Doing so introduced a powerful idea of "objects"—chunks of code so fundamental that they could be transported from one program to another virtually unchanged. Smaller modules also made it easier to isolate and debug problems, were they to occur once the program was up and running. For instance, a Spy program might have a module dealing with processing a request for help.

```
500        !
510        !
980        !Help:
990        !
1000       OPEN "INSTR.TXT" AS FILE 1
1010       FOR I=1 TO 500
1011       INPUT LINE #1,A$
1012       A$=CVT$$(A$,4%)
1013       PRINT A$
1014       NEXT I
1015       PRINT #2,"HELP"
1020       GOTO 20
1070       !
1080       !
```

The help module, identified by exclamation points (which tell the computer to ignore that line, as it is meant to be read as a "comment" from one programmer to another within the code), sits between other modules with names like "Delete" (for removing a spy from the database), "Records" (the module for changing a line of data from the database), or "Search" (which finds an agent in the database). Such attention to neatness was rewarded by Mr. Moran.

While working on the innards of the code was tremendously satisfying, the real glory came from conceptualizing and adding new functions to the standard Spy program. These upgrades were of our own imagination. The challenge was to pick one that was neither too frivolous nor so complex that it ruined the rest of the program. After five years of students writing Spy programs, it was difficult to imagine what new functions could possibly be added. I decided to focus instead on supreme modularity, elegance of internal code, and on the "front end"—the user interface—producing clean, functional commands and a help system that made sense.

One day in the late winter of tenth grade, while working on internal comments to go within the code, I began to doubt my decision. The evening sky turning to black, I felt warm and comfortable by my terminal, surrounded by printouts, the fluorescents casting a bright glow on the green-and-white paper. Across the table sat Boz, working on his Spy program. He'd been up to something, I could tell from his unusual silence. *What is he working on?* I wondered. Boz stood up and went over to the printer as it clattered to life, tearing out a page when it was done. "Yes!" he said, "it works!"

Too proud to jump up and say "What?" I ignored Boz and kept coding. "Mr. Moran," he said. "Look at this."

He ran up to Mr. Moran's desk and spread the printout before him. I couldn't help but overhear what was being said, and I thought, *He did it! He actually found something new to do that no one's ever done.*

Boz held an airline ticket, printed out by his Spy program, to send an agent on assignment. Everything was in place: departure date, return date, cities, fare. He'd created a whole set of data tables so the spymaster could not only assign an agent but also book him out to his destination! This required complex instructions to open the printer as a "slave"—a subservient device to his Spy program. These instructions, however, also could be used in theory to open another terminal—say Mr. Moran's—as a slave, thereby producing a security violation. Haahr had originally hacked out a brilliant program to give students control of the printer, without a backdoor to controlling terminals. Boz had pulled elements of code out of Haahr's program and re-tooled it within the parameters of his Spy program. I stood up to see what Boz had done. The ticket was even nicely formatted. It had a border around it, and the text was centered like a real ticket. I was amazed. "How did you do that?" I asked. And in keeping with the unwritten ethic of the room, Boz told me.

Misha wasn't around that afternoon. When he heard of Boz's neat hack the next day, he steamed with fury. Complaining to me, he derided Boz's airline ticket module as a flashy stunt. Misha was particularly intense and competitive when it came to programming. With no other extracurriculars to speak of beyond the room, events there took on particular weight for Misha. I told him I thought Boz deserved credit for what he'd done. Over the year Misha, Boz, and I had formed an odd triangle, with me in the middle. Misha despised Boz. Misha's obsessiveness and need to track what each of us did with our programs fit perfectly with Boz's desire to be noticed. Boz didn't like Misha either. Their dislike bonded one to the other—Boz at times overtly showing off in front of Misha, Misha rising to the bait, waiting for the showing off to begin. Misha would sit quietly at his terminal, deep into the coding, then suddenly look up with a furious expression if Boz started showing off. Misha's brooding silence would push Boz to greater acts of extroversion—louder boasts—which in turn would draw Misha's interest.

When we handed in our Spy programs in the early months of 1984 I knew that the three of us were all good, and that at the end of the day, Misha actually was the best programmer among us. The summer before, while I'd worked with my stepfather's Microdata, Misha had earned thousands programming down on Wall Street. He'd continued working through school and had been exposed to languages and problems none of us knew. Rather than focusing on who would get the highest grade, I found myself drawn out of the room and into the cafeteria more and more frequently. There I could flirt, hang out with girls, and pursue what was fast becoming a primary objective: having a girlfriend. I hadn't had one since Jennifer in seventh grade, and as I turned sixteen I bemoaned my predicament: I might graduate from high school a virgin! And sitting in the computer room would not help.

A week later Mr. Moran returned our programs. On the top of mine Mr. Moran had written "A." I was extremely proud. With an A in Computer II my entire grade-point average would rise to a B, a minute fraction away from B +. I'd never had a grade-point average that high in all my life. Misha and Boz also received A's. We'd completed our time with BASIC. For the remainder of the year, Mr. Moran promised a survey of other languages—Ada, LISP, COBOL, FORTRAN, ALGOL. We'd spend a few weeks on each, learning the syntax, hacking out simple programs. The idea was to expose us to the multiplicity of computer languages and understand how each one had particular strengths. COBOL was excellent at maintaining databases. FORTRAN was best at executing mathematical formulas. LISP excelled at creating abstract "sets," best used in artificial intelligence programming. For most in Computer II, this would be their last computer class, and Mr. Moran wanted them to leave able to take on computer science, if they chose, at college. A few of us would stay on for a new course: Advanced Placement computer science. We would be the second class in the history of high schools across America to take a nationwide computer science exam, one that could provide successful students (who received either a 4 or 5 out of 5 points on the test) with credit for a first-year, college-level computer science course.

Throughout Computer II I'd watched as Haahr and Hilal completed assignments for AP computer science. They were the first to take this course, which had been approved by the College Board only the year before. The board had decreed that the AP exam would be given in Pascal, because Pascal was both a "universal" language—it tended to be similar from computer system to computer system—and its structure and syntax reinforced good programming habits. Where BASIC tempted programmers with analytic shortcuts, most notably overuse of the GOTO command to "escape" from tangled code, Pascal by its nature reinforced modular programming and the creation of descriptive variables that another programmer reading your code could understand. Mr. Moran had offered the AP class as soon as the College Board approved the curriculum. Providing an AP class moved the computer room one step closer to acquiring the legitimacy of English or physics.

The first inkling I had of an alternative to the BASIC way of thinking came from Paul Haahr and Paul Hilal. Their programs hinted at symbolic structures I hadn't imagined. In the spring of 1984 I would come into the computer room and find Paul or Paul printing out Pascal programs. They were hard at work writing an elaborate program we called the Cheese program. Cheese was a special, multimonth-long assignment that Mr. Moran had devised to teach every possible principle of good computer programming using Pascal. Only students in AP computer science wrote Cheese programs. The Pauls were forever working on their versions, adding yet another function, discovering some new way to devise data sets and sort their inventory. Meant to simulate the workings of a cheese factory, the program had to handle inventory control, managing piles of data on what cheeses were in or out of stock—gouda, havarti, brie, Swiss, and so on—and reorder what was missing. Most difficult of all were the reporting functions. The program had to provide statistics on which cheeses sold best, what combinations of cheese shipped most frequently, and financial averages on the cost of typical cheese orders. These then had to be printed out in neat reports, the kind a factory manager would like to read.

Cheese permeated the room—cheese, cheese, cheese. Scraps of sorting functions with telltale variables like "Muenster" or "cheddar" would be left half erased on the board. Printouts, sometimes shredded, other times left underfoot, would clutter the area around the printer. Picking them up would reveal—cheese. I began to feel that Cheese was all that mattered—and that Pascal was the language I had to learn. But Pascal, from what I could tell, was a complete mind bender. It made no sense. The seduction started the way seductions always do—with good looks. Pascal is beautiful, I thought. Pascal doesn't use line numbers! 50 GOTO 30 doesn't exist! How could a computer use a programming language without line numbers? In BASIC everything starts with a number. How did the computer go down the program without numbers, or jump around, as in GOTO 200? Whereas our BASIC programs were always long-running strips of code, Pascal programs wove in arcs . . .

XXXXX
XXXXX
  XXXXX

```
XXXXX
    XXXXX
    XXXXX
        XXXXX
    XXXXX
XXXXX
XXXXX
```

Pascal, I began to sense, was the best, the purest language. Freed of numbers, code seemed to float within the empty space of the white page. Pascal programs did not read from the top down, they read from the inside out. Pascal held things within things, which could be within something else. In Pascal single words can signify entire programs within themselves, which in turn can be embedded within another program, the two of which could be given a name of their own and placed inside a third! The power! The simplicity! How to keep track of so many levels, or make order from these strange symmetries? Picking up scraps of the Pauls' programs—either of the Pauls—I was mesmerized by the words on the paper. One evening Paul Hilal printed his Cheese program and I asked him if I could look at it. He gave it to me, and I took it home.

I spread the pages out on my floor and started at the top.

```
PROGRAM Cheese (input, output);
```

Spreading out below was code like none other I'd ever seen. It was so beautiful. And so hard to follow. He had these marvelous variables of TYPE cheese. By creating a TYPE, Paul could dictate the essence of a variable. In BASIC a variable could equal a number or a set of characters, but in Pascal variables could be of completely new types. Instead of having just two types of variables—numbers, like 1.347 or –10, and characters, like "yes" or "a long phrase"— you could create TYPES of, say, cheese. The TYPE cheese could be defined as gouda, havarti, brie, cheddar, Swiss. That means a variable of TYPE cheese, say the variable cheeseorder, could be defined by stating:

```
VAR
  cheeseorder : cheese
```

Later you could command IF cheeseorder = brie THEN DO something where "something" was a set of commands. The great thing about types is that you can never assign cheeseorder a value that isn't one of its type. Cheeseorder : = 252 would return an error. So would cheeseorder : = "brie" because that's the word *brie*, not the TYPE cheese of kind brie. This language was elegant. It had lovely details. All the commands in his program appeared in UPPERCASE, creating a visual flow. And the lines were indented as the program got deeper and deeper into what were called "procedures."

But I couldn't learn Pascal yet—Mr. Moran was saving it for AP computer science. I went downtown anyway, to a bookstore on 51st street and Broadway that sold computer books. There I bought a precious book, the same one Mr. Moran used—*Programming in Pascal*, by Peter Grogono. It was dense and typographically curious, written in typewriter print. All the other programming books I'd seen were written for specific machines, like my book *Atari BASIC*, but this book was for a kind of abstract computer, a universal machine. That's part of what made Pascal so exotic. In the balkanized world of computers, where a code for an IBM-

PC could not work on an Apple, the idea of a universal language—a digital lingua franca—had enormous appeal. Pascal was meant to be that. Reading a Pascal program was like reading poetry, a poetry of logics and esthetics that took me far away from the world around me. It brought me closer to the computer as an artistic medium.

When Paul and Paul graduated in June, the half dozen students who enrolled in AP computer science became heirs apparent. Strangely, though, the chain had broken. When we would come back as juniors, there would be no seniors in AP computer science with us, and for the first time, at the end of the school year Mr. Moran didn't appoint the next year's Super Users. We would return the following fall to a room without one.

10

Cheese

ONE PROCEDURE KEPT CRASHING my AP computer science
Cheese program. It seemed simple, a routine control structure. The procedure read from the database, looking for a
particular cheese. The idea was to scan each record, checking if it equaled the cheese value sought. If it found a
match, the procedure copied that record's address and returned to the previous part of the program. Mission accomplished. If, however, the cheese was not there, the
procedure tried again, moving down to the next record. This
was a generic conditional loop, designed to persist until the
desired result is attained. But for some reason it didn't
work. If the chosen cheese didn't come up in the first dozen
or so records, the program fatally crashed, claiming it had
run out of memory. That made no sense. It was just a loop.
The procedure looked like this:

```
PROCEDURE get_cheese(VAR wanted_cheese :
cheese, output_location : integer);
VAR
  cheese : cheese
  location : integer
```

```
BEGIN {procedure get_cheese}
 read_database(cheese, location)
 IF cheese = wanted_cheese
  THEN
    output_location := location
    END
     ELSE
       get_cheese(wanted_cheese, location)
END; {procedure get_cheese}
```

The idea was to scan the database of cheese by calling the get_cheese procedure. Get_cheese goes through the database using another procedure called read_database, which I created to handle all data search requests throughout the entire Cheese program. That's the power of Pascal: you write one procedure and can use it throughout the rest of the program rather than rewriting the command over and over. If read_database does not find a cheese value that matches what you want, get_cheese moves down to the next record by calling itself. I pictured the whole thing like a GOTO loop. The little pointer moves down the procedure, and if read_database does not return what is sought, then get_cheese invokes get_cheese(wanted_cheese; location) and starts over, until the wanted cheese is found.

But then it crashed.

```
[PROGRAM STOPPED OUT OF MEMORY]
$
```

Why?

That fall in 1984 of my junior year, as I'd thrown myself into the heady realm of hacking cheese, I experienced a growing dislocation at home. I lived in two separate worlds, compartmentalized, kept apart. Where in my early teens I'd accepted my parents' divorce and their new marriages, as my seventeenth birthday approached pent-up anger and disdain for my parents, stepmother, and stepfather welled up. I'd rarely share my life at school with them, speaking little of my friends, classes, or the daily events of hanging out in the cafeteria. Much as I'd prevented my father from sharing my pleasure the day he first saw me program on the Atari, I'd fallen into a persistent pattern of putting a wall between me and them. In part the anger came from fear, a sense of things gyrating out of control as my sister began to test limits, pushing beyond to a place I'd never dared to go. Where at thirteen I'd veered away from one path with Eric and Tim, my sister, now that age and in eighth grade, hurtled forward, seemingly unstoppable. I'd found a compelling alternative in the gift of a computer; what might the equivalent muse be for Samantha? My parents agonized over this, and the strain of separation, of families broken and rebuilt, was compounded by the awful realization that she was set on growing up fastest of all.

As I focused on my Cheese program and life at school, I didn't see much of Samantha. Although a few months earlier my father had converted the den into a room for her, so we could live together instead of taking turns at each parent's house, our worlds had diverged in a matter of one year. After her expulsion from Hewitt at the end of seventh grade, the local public school, Robert F. Wagner, had taken her in. There my sister skipped class and instead hung out in Central Park where groups of kids from schools all over the city would meet. My mom and dad had no idea she was missing so much school; her teachers never told them. Samantha liked to spend afternoons in Sheep Meadow, on the lower end of Central Park, which served as a communal high school drug bazaar. She'd fallen in love with a man twice her age, a twenty-seven-year-old pot dealer who worked the Meadow.

Just a bit taller than my sister, maybe five-foot seven, and skinny, Brian was built like a Heavy Metal rocker and had a pasty, boyish face showing stubble in soft patches, and un-fashionable, hippie-ish blond matted-spaghetti hair. His dad was a lawyer, and Brian was squatting in his father's client's apartment on the East Side while the client had temporarily moved to Europe. Samantha was slowly, inexorably, run-ning away from home to be with Brian. First by skipping school, later by staying out late on school nights, and even-tually not returning home for a day or two at a time. My parents chose to negotiate, setting flexible rules—she could go out on weeknights if she finished her homework, but had to be home by midnight—yet, step by step, the rules were broken, renegotiated, grudgingly extended. Lurking was the unspoken fear that if we pushed too hard she might never come home. I was the only kid in my grade who had a semi-runaway sister, a thirteen-year-old who did not al-ways come home to sleep, a girl who spent her nighttime in houses and apartments in places I didn't know. Little sisters aren't supposed to do that. And big brothers are supposed to protect them.

*   *   *

Grappling with my program at school, I knew the get_cheese variable couldn't be the problem. It had worked flawlessly in other procedures throughout my Cheese pro-gram, so at school I loaded the DEBUG program that comes with the Pascal compiler. DEBUG let me step through the code, one execution at a time, watching exactly what the PDP was doing. That time, DEBUG didn't help. I watched as the program stepped, one instruction at a time, just as I thought it should: moving down the procedure and calling itself until the cheese is found. But every time, around the twelfth iteration of the loop, the program crashed. Why, why, why? It turned into one of those ghost-in-the-machine moments, when I'm convinced computers are much more mysterious than I realize. Perhaps sentient.

Boz and Misha were in the room too, across from me on the other side of the table. I was embarrassed to ask Mr. Moran for help in front of them. What if it's a stupid problem? So I ran DEBUG again. Maybe I missed something.

```
[PROGRAM STOPPED OUT OF MEMORY]
$
```

Mr. Moran sat as his terminal, at the head of the table. I could just ask him, but . . . "Mr. Moran," I said quietly from my terminal near his desk, "Can you help me?"

He looked up and told me to come over. I pulled up a chair beside his desk and laid out the program; we could look at it together.

"This procedure is crashing, and I don't know why. See, this is what it's trying to do." I explained it to him. Mr. Moran used his silver pen as a pointer, just as I imagined the computer did, somewhere in the abstract matrix of memory. He followed the lines, pointing.

"Hmmm," he said.

I felt better. Hmmm was good. That meant my question wasn't so stupid after all.

"Hmmm. It looks right," he said.

I was thrilled. Mine was a smart problem.

"Let me see," he said, getting up. We went to my terminal and I ran DEBUG, taking him to the crash. I pictured the magic pointer moving up and down, like a finger passing along a stack of books.

Mr. Moran looked at the code.

"Of course," he said, suddenly smiling, his face turning red.

"This isn't a loop. It's recursive. Every time you call get_cheese it calls itself inside itself. If the condition is still false, it calls itself again, until the computer runs out of memory."

I was confused. "It calls itself inside itself?" Inside itself? Then it happened, as if the floor fell away from my feet and I too was falling, the bigness of the whole thing suddenly in my stomach. This isn't a loop! It's a snake eating its tail! It's infinity, the procedure self-replicating inside itself! It's creating a whole new universe inside another universe, and again and again, and would do so forever, were it not for the limits of the PDP. One inside the next, except each one is both the same size yet inside the preceding one, an impossible simultaneous state of two existences. Of course the PDP crashed. Nothing finite can contain the infinite. I'd never felt that before, the reality of infinity. There it was, a wordless revelation.

"Thanks Mr. Moran," I said, and he went back to his desk.

On the way to school, riding the Number 1 uptown, I'd unfold the printout of my Cheese program out on the subway seats. I'd become enamored of data tables. They were like a coal mine, rich deposits of information that could be processed and organized in myriad ways. Thinking about data was comforting, and challenging. Our Cheese programs, while simulating the business of manufacturing and distributing cheese, demanded an extension of abstraction beyond the code of the program, into the data it both consumed and created. How should the data be organized? What forms were best? Because the computer could run through and produce entirely new tables of data in a matter of seconds, organizing information was an integral part of a successful Cheese program. The best form for this problem, as I learned from Mr. Moran and through experimentation, was something called an "array." My book on programming in Pascal helped me.

```
CONST
  maxcol = 80;
TYPE
  colindex = 1..maxcol;
  cardimage = PACKED ARRAY [colindex] OF
  char;
VAR
  data : FILE OF cardimage;
  inputcard : cardimage;
```

I was having a difficult time working with data files. It was hard to understand where one record ended and another began. As my array of cheese inventory expanded, comparing records and retrieving them required elegant solutions rather than brute force tracking. Mr. Moran helped me by introducing the idea of keys. Keys are special markers in the records that I put there on purpose; these were easily detected by my program. By tracking keys I could tell where I was in the big array of records. This allowed me to jumble the data, sending it topsy-turvy, knowing that if need be I could restore it by ordering the keys. Records could be reassembled and the great fear of losing their constituent parts avoided. Using keys, I could rearrange the entire topography of my data. With one command it would reorganize along various lines—alphabetical, reverse-alphabetical, number in stock, out of stock and reorder date, out of stock and reorder date alphabetical.

One afternoon Boz and I listened carefully as Ian explained how he'd stored his cheese in a binary tree rather than a packed array. A binary tree resembles the root system of a plant: a cascading series of points, each of which produces two lines terminating at two points, which in turn sprout more lines. These points are called nodes. Boz and I had organized our data in a two-dimensional matrix, like a series of letterboxes in the lobby of a hotel, but Ian's data was an organically shaped pyramid. I had difficulty grasping the structure, which was much more complicated than a rectangular matrix. Ian, however, had no trouble at all, and his ordering of things allowed for something special: lightning-fast sorts, and the ability to "grow" data downward without having to track the exact size of the data table. Mr. Moran had told us to experiment with storing the cheese in a binary tree, and Ian was the only one of us who understood how to delete cheeses from the tree without losing the overall order of all the cheeses. He used one procedure, calling itself recursively. Pure coding, that was.

AP computer science made us feel elite. We grasped ideas most people would never learn, concepts that in a world measured, monitored, and controlled by computers allowed us to understand everyday systems. I found myself staring at cash registers, noticing how computer monitors at checkout counters are used to restock products before they run out on the shelves. At Radio Shack, which always asked for your name and address, even if you paid in cash, I admired how much data they were collecting every time I went in there to buy more disks for my Atari. I wanted to play with arrays of that size—millions and millions of records, each denoting a real person, sifting through to find unexpected patterns. Perhaps animating the data, finding visual shapes that otherwise would escape the eye as it scanned a long paper printout littered with static numbers.

Mr. Moran treated us differently as we grew older. He began to see us more and more as intellectual peers and as people he could have fun with, too. On Friday afternoons in the last fifteen minutes of class he would challenge us to a classwide game of BLOCKADE. It was like SURROUND on the Atari. Both were based on the scene in *Tron* when the hero rode a "light cycle" that left a trail of phosphorescence hard as a rock. The goal was to box your opponent within the immobile exhaust, and since the bikes had no brakes, you'd eventually crash and die. BLOCKADE on the PDP left a trail of Xs instead of a three-dimensional sky-blue wall. For the PDP to draw these Xs simultaneously on our monitors took all the capacity of that machine, and Mr. Moran would temporarily suspend every job on the system, even BIGBRO and LILBRO, so that nothing would interrupt our game.

One Friday Boz and I found ourselves winning as one by one the rest of the class crashed into our trails, each impact a point. Ian valiantly tried to survive within a box I'd cast around him by following its inner wall as closely as possible then executing perfect 180-degree turns one row lower, hoping that his own boxy inward crushing spiral would outlast whatever obstacles we on the outside might encounter. I tried to box in Boz but he slipped out and turned around, suddenly on the outside moving to lock me in. As we raced to the far wall, I saw another trail suddenly cutting down from the top of the screen. Mr. Moran. As if we were one, Boz and I turned in unison and headed for the teacher, closing in on two far sides. We raced to the far wall, pursuing Mr. Moran, our Xs blazing, the imaginary cycles in our minds humming, converging. We crossed paths turning together at the last instant, side by side, closing off Mr. Moran's exit. As we spun in a parallel inside spiral, with Mr. Moran on the innermost loop, we knew we had our quarry. But Mr. Moran would push us to the end. He headed for Ian's wall, forcing us to move down faster, cutting down our time, our space. My fingers clung to the keys. We held our rhythm and drew up along Mr. Moran, the three of us headed for Ian's trail until, at the last moment, Boz and I peeled off to either side and Mr. Moran had nowhere to go but headlong into the wall before him. Later Boz and I highfived when Mr. Moran wasn't looking.

Life was good. I had Cheese, I had my books, I had a refuge from home, with the bits in the system flying, flying in those shapes that came up in my head when I programmed. At night when I worked on my Cheese program Samantha wasn't there. She had turned into a skillful liar, and my parents were slow to realize she wasn't doing anything to get better. It was an impossible situation for all of us. I worried she might never get better. Every time they gave her a choice, a way out, Samantha would turn inside, the spiral getting tighter, the way out smaller.

"I'm going to have a party at Papa's," my sister told me one afternoon. That's what we called our dad, "Papa," from our time in France. He was going away on a business trip for a week and the apartment would be empty. The thought of Samantha bringing her druggie friends over to our house made me nervous. What if they stole my things or wrecked the place? "You have to be careful," I told her. "I mean you can't let them take things, you really can't. And no matter what, they can't go into my room."

The night of the party I went to my mom's house, figuring I'd spend the weekend there and avoid the situation. My mom had no idea Samantha was throwing a party at my dad's. Even though they lived ten blocks away from each other, it was easy to keep the two of them in the dark. But at 1 A.M., unable to sleep, bored with television, anxiously wondering what might be happening, I decided to quietly slip out of bed and go to my dad's. He lived on the twentieth floor of a Park Avenue building, and as I entered the lobby the doorman gave me a look that said, Do you know what's going on up there? I pretended not to notice. As I stepped out of the elevator I could hear music, voices screaming, laughing, talking, singing. The faint sweet smell of dope drifted out from under the apartment door.

God I wish this was someone else's house, I thought as I walked inside. And someone else's sister. Then I could have a drink, maybe smoke some pot, and meet a pretty girl. But this is my home, and my sister, and these are her friends. And it's out of control. Five girls in black jeans were smoking a joint by the door. A pile of guys were doing coke on our kitchen table. People were dancing in the living room. The ceiling was coated in a misty blue fume of smoke and sweat and pheromones. A million plastic cups and cigarettes seemed poised to land on the rug, burn the couch, stain the table. I didn't see my sister right away. Then I found her, in the kitchen with the cokeheads. Brian was there. He laid a come-on-man nice guy hustler smile on me. I wanted to take one of the big knives from the cutting board next to the oven and shove it down his throat.

"I'm going back to mom's," I told my sister. Brian smiled and nodded and waved at me.

The next morning I came back. The house was freezing. All the windows were open, and a strong wind came through, rattling the plastic cups and cigarette butts on the floor. A chair was tipped over on its side. My dad's bedroom door was closed, and I opened it. Samantha and Brian were asleep on the queen-size bed, naked, curled around each other. On the floor was a big mirror streaked with white dust, and the wind had blown what was left on the carpet, forming a faint pattern that stood out in the gray light. I shut the door quietly. In my room I found the computer as I'd left it. It was safe. Samantha had kept her word.

*   *   *

I got an A on the first part of my Cheese program. So did Boz, Misha, and Ian. The competitiveness was building. We were immersed in a tactical conspiracy, a network of shifting alliances similar to tactics I'd used years ago in my strategy war games with Jesse. Misha and I would call each other on the phone at night to talk about Boz and Ian. Ian called me to talk about Misha and Boz. Boz rarely called me. But sometimes in the computer room, when the others were not there, Boz and I talked about who would become Super User. Misha called me the most. His obsessive desire to beat Boz fascinated and repelled me. While Boz could be brazen and loud, he was a good kid, wearing his intentions on the outside for all to see. But Misha was all secrets, hidden intentions, intentions within intentions. He was unpredictable to all but himself. To most his demeanor remained one of brooding silence, with a simmering intensity that would burst through to the surface unexpectedly. Misha, Boz, and I were the most likely candidates for Super User. We were the best programmers. And we were more enthusiastic about running and maintaining the systems in the room than anyone else in our class. Perhaps we would all make it. We could share the title together.

My father found out about the party. Although Samantha managed to clean the apartment, someone had burned a brown circle in the white lacquered top of our dining room table. My dad was furious, but what could he do? Once a child stops caring about parental approval, little leverage remains. Samantha's orbit was shifting toward Brian, the mass of his approval drawing her out of our system and into his. Brian most of all made her situation totally unlike the one I'd faced at her age. But my parents rallied. My mother and stepfather bought a new apartment, one large enough for my sister and me to have our own bedrooms. Perhaps by having us together, neither favored over the other—my sister wouldn't be sleeping in a den but in a room of her own—stability would return, blunting Brian's power.

In early 1985 Samantha and I moved into our new home. My room was toward the back of the building, a ten-story apartment house on 77th Street. Samantha's was on the other side of the apartment. Our mom had spent months fixing up the place, making it perfect for all of us, each detail a tangible expression of her love. It quickly became apparent, however, that Samantha would persist in her ways. At night when she was out I'd sit at my desk doing homework. My small, comfortable room looked out on the courtyard formed by the back of our building and the back of a building across the way. The windows on the left side of the courtyard were ours. If I tilted my head I could see Samantha's room. I was more interested in looking straight ahead at the apartments facing me. Working on my Atari, I would stare out the window at all the families having dinner in their kitchens, each a little television screen, a silent show. I loved this view. Sometimes I'd turn off the lights and sit in a chair by the window so I could watch without being seen, especially when the girl across the way was in her kitchen.

She was around my age, sixteen or seventeen, with brown hair in a pageboy cut, big breasts, and legs like a swimmer's. Her kitchen was on the fifth floor and I was on the seventh, so I could look down at her. When I was doing my homework I would look out every few minutes, checking if she was in the kitchen yet. Sometimes she would come in wearing a long white T-shirt and her underwear. *Does she know I'm watching?* I wondered. I wanted her to know, but at the same time I was afraid it would end if she did.

One night she entered the kitchen wearing only a white bra and panties. She sat at the kitchen table and casually opened a notebook and a textbook. She read. I stared. I switched off the light in my bedroom and waited, without moving, for my eyes to adjust. Then slowly I moved my chair to the window and quietly lowered the venetian blinds, tilting them open so I could see between the slats. She was in profile. Her leg jounced up and down. Her pen tapped the table. She was dancing to music. Then she stood up. Exercising? Dancing? What did it matter? Suddenly I realized she knew I was watching. She is doing this for me. Our secret flirtation, eyes averted, never looking directly at each other, continued night by night like that. Every now and then I'd look to my left, to Samantha's room, which was almost always dark, a black square, nearly invisible, on a dark wall.

Soon after I moved in with mom my American history teacher, Mr. Huber, told us that we had to write a twenty-five-page term paper. The paper would be typed, not hand-written. I decided to write a report on the history and importance of space exploration. I wasn't sure what was better: reading books on outer space for school credit or using my word processor. My copy of Atari Writer let me type into my computer and edit the document. I also had an Epson dot-matrix printer, which used a series of tiny pins to produce letters. Mr. Huber hated dot-matrix printouts. He said that anyone who handed in a paper on dot-matrix would automatically go down half a grade. If I wanted to use Atari Writer I'd be in big trouble.

There was more to this than dot-matrix printers. I suspected that Mr. Huber actually *hated* computers. Worse, he didn't trust them. He believed that writing on a typewriter forced us to think better. But he wouldn't come out and say, "don't use a computer." *He's pussyfooting around*, I thought, coming up with dot-matrix excuses. I would not write my paper on a typewriter! No way. I had a plan.

In the computer room we had a letter-quality printer, our Diablo, which enchanted us with its perfection. Until its arrival, our only printer had been a 24-inch accounting printer that used long sheets of white-and-green ledger paper with sprocket holes on either side—fast, dot-matrix. Were I to hand in a paper using that printer, Mr. Huber would probably mark me down a full grade. Death by dot-matrix. But if I could print out my paper on the Diablo, Mr. Huber would never know because it used daisy wheels to print. The daisy wheel was a black piece of plastic with letters on it. By inserting different wheels you could produce different typefaces. Mr. Moran called these "fonts." If I used the classic typewriter font and printed on cream-colored paper, it would seem as if I'd labored for hours to produce an immaculate copy. This was one of the ancillary benefits of the powerful text-processing abilities that were appearing in the mid-1980s: the image of perfection, at a fraction of the time. Or so they promised.

I would write my paper at home on the Atari and then upload it right into the PDP's word processor using my modem. At 1,200 bits per second, which amounted to 120 characters per second or, by my estimation, 24 words per second, my paper would fly into the system at superhuman speed. Then, first thing the next morning, I planned to print out my paper on the Diablo at school and hand it to Mr. Huber in class that afternoon.

I'd spent several weeks researching my paper, and after a week of writing managed to finish at seven the night before it was due. An informal telephone network of friends calling friends told me that a good part of our class would be up all night with their typewriters, scrambling to finish. I figured I'd be watching TV by eight. My paper consumed 60K of disk space, which by my calculation would be equivalent to twenty-seven typewritten pages. At a time when what appeared on screen bore little resemblance to how it would look on the printed page, I'd memorized clever formulas to translate disk space into words and pages. Outside it was dark, and by the glow of my blue television screen I flipped on the modem, dialing Horace Mann. The PDP welcomed me with its atonal whistle and I logged in, entered the PDP's word processor, and downloaded my paper.

34^^&^%TfdTYHGdf^$%G#CF$TYQ%GRQ34yQgAgy hawq35 v45y* is all I saw of my paper on the screen.

Twelve hundred bits per second was too fast. I slowed down to 300. Still too fast, though now I could make out fragments of what I'd written. T#%h(e wor@(#)%ds look*d li*#"ke this. The PDP couldn't keep up. I brought it down to 110 bits per second, the slowest setting on my modem. A marginal improvement. Instead of every word coming up as gibberish characters, only a third of them turned out that way. I resigned myself to fixing it manually. By 5 A.M. I'd finished, essentially retyping the whole paper online. The blue light of dawn arrived, slowly overtaking the blue light of the monitor. I'd pulled an all-nighter, not writing, but hacking. It would work.

"The printer is broken."

"What!"

"Yes, it's broken. There's something wrong with it."

Boz grinned as he told me this. It was 8:30 in the morning, and we were in the computer room.

"Mr. Moran has to order a new part."

I couldn't believe it. Late papers were unacceptable. We'd be marked down a full grade for every passing day.

When Mr. Huber castigated me in front of the whole class, holding up my dot-matrix printout on 24-inch paper, I was heartbroken. My excitement in trying to use the computer to produce my paper was not only the joy of the hack; it was the appeal of using the machine to solve a real-world problem, especially one far removed from algorithms and theoretical cheese factories: to produce a beautiful paper that didn't look like I'd used a computer to write it. Using the machine for an artistic end, a non-technological goal, was a route I wanted to explore after high school. In college I pictured myself doing something exciting with computers that was hard for me to articulate then, but I had glimmerings of uses that were more, well, *human*, integrated with the whole of life rather than compartmentalized in a special room. My history paper had given me a taste of that, bitter as the experience was. When Mr. Huber gave my paper back to me, my B+ had, as he'd warned, been lowered to B.

Not long after my dot-matrix debacle I was walking home from school when I saw her coming toward me. We stopped in the median of the street and stared. "How do I know you?" she asked.

"I don't know," I said. She was familiar. Maybe we'd met at a party. Pretty, but a bit heavy. I recognized her from somewhere. The window?

She smiled at the same time. Feeling embarrassed, unsure of what to do, before I could think further she spoke.

"From across the street," she said.

"Yes," I replied, and we fell into talking. I walked her to the drugstore, where she was headed. "That's your building," I said, pointing to it.

"Yes," she said, and later as I watched her buy shampoo she asked if I'd like to come over and see.

Standing in her kitchen looking up at my window, we laughed. I could see my room, the side of my desk. Her name was Wendy, and she went to Riverdale. We knew many of the same people. I spent a while with her, drinking soda, talking, and from then on it was never the same. She had a name now. We would wave to each other through our windows, and she stopped dancing late at night.

In the spring of 1985 as the school year was finishing, I began studying in earnest for the AP computer science exam. After the AP physics exam, AP computer was regarded as the hardest test given to high school students. I would have to work for my grade. So I was surprised one afternoon when Boz came to me in the cafeteria with an extraordinary admission, one that would put everything he'd worked for—as well as me—in jeopardy.

"I got the passwords for the 186 group." Boz announced.

That was our grade—the 186 group. Boz had my password. He had Misha's. A cardinal sin. Stealing passwords.

I looked around to see if anyone was listening. No one was.

"How did you get them?" I asked.

"The terminal was unoccupied," Boz said, meaning Mr. Moran's, "and I typed SYSTAT and saw it was a Privileged Account. I just had to look." He glanced around the cafeteria and then he opened his bookbag, revealing the printout.

"Boz, you're crazy. If Mr. Moran finds out—"

"He won't."

"Have you told anybody else?"

"Misha."

That he'd shown it to Misha was utterly strange. Misha would surely use this against Boz. But Boz was like that—for all his braggadocio, he was trusting. Boz said he was going to destroy the printout.

Later that night at home I thought about it. The rules were clear, stealing passwords was unacceptable. And no one who wanted to be Super User should do it—least of all at this stage, at the very end when we were so close, when we had proved so much to Mr. Moran. Boz had turned me into an accomplice by telling me. I picked up the phone and called Misha. Immediately, he told me what Boz had done. I told him I knew.

"I think we should tell Mr. Moran," Misha said.

"I don't. Boz might get expelled, you know."

Misha and I argued about what to do. He was absolutely firm. If we wanted to be Super Users we could not be lenient about basic security. Boz hadn't misused the information. He hadn't deleted any files or done anything destructive. He merely had gained knowledge that was forbidden. I was torn between Misha's logic and feelings of forgiveness. Curiosity. Temptation. Forgiveness.

"I'm going to tell Mr. Moran," Misha said.

"Yes," I agreed. "We should tell Mr. Moran."

A cursed mission. Boz would surely not become Super User. I vowed to fight for leniency; I would tell Mr. Moran that Boz seemed penitent, that his telling me was a sign of his desire for forgiveness. He was a good programmer, and I knew, much as we'd competed, that I'd miss him, that the room would seem empty without him there. Had I been in the room alone, trusted to watch over the terminals, and had I then found a terminal with Super User access unattended, I too might have looked up the password file. Even if I hadn't printed it out. Surely I would have looked at the screen, leaving no trail, a fleeting glance at the forbidden.

That morning Misha and I went to see Mr. Moran during first period. On our way down the hall I was having second thoughts. "Are you sure we should do this?"

"Come on!" Misha said. "He took our passwords."

I wanted Misha to tell him. Maybe he would tell him without me. We went into the system room and Mr. Moran was standing by KB:0, the master terminal, typing.

"Good morning," he said, looking up.

"Hi Mr. Moran," I said. "We have something to tell you."

I knew then that I would say it. I wanted to. Perhaps then I could modulate what had happened and by coming forward have the moral authority to draw leniency from Mr. Moran.

"Jeremy Bozza hacked into the eleventh grade password file and got all our passwords."

Mr. Moran blinked once or twice; other than that he seemed implacable.

"Are you sure?"

"Yes," I said. Misha nodded with me.

"How do you know?"

"He told us," Misha said.

"Did he show you the file?" Mr. Moran asked.

"Yes," I replied. "He did."

For a long while Mr. Moran was silent. He had a very determined look on his face. I could tell he was extremely angry.

"He's going to be all right?" I pressed. "Right? I mean, nothing bad happened. He didn't break into anyone's account." I was sure this was true.

Mr. Moran said only, "I will speak to him." I went downstairs to the cafeteria by myself and bought a bagel and orange juice, and stared out the wall-sized plate glass windows at the football field, which was green with spring grass. A few kids lolled around on the lawn, doing homework together.

Later that afternoon I went to the room and found Misha working at a terminal. I sat down beside him. Before logging in I quietly said, "Have you seen Boz yet?"

"No."

Just then Boz walked in the door.

"Hi," he said to us, all flushed. A new period was starting, and Boz had raced up the stairs to get here. A few seventh-graders in Beginner BASIC straggled in. We ignored them.

"Hi," Misha and I both said.

Boz sat down to log in. His fingers quickly tapped out the commands. We could see Mr. Moran coming to the door and opening it. He stepped into the room.

"Jeremy," he said.

Boz turned around in his chair.

"Can you come in for a minute, please?"

Boz typed for a few seconds—probably initiating OCCUPY so we couldn't peek at his account while he was away.

Boz stood up and went toward Mr. Moran.

"In here," he said, indicating the system room. They went in and Mr. Moran closed the door.

Misha and I looked down at our screens. I slouched down a bit as if I could disappear, slip into the screen. I looked up. The system room was fairly soundproof. I felt as if I was in *2001* when the two astronauts sit in the pod and HAL lip-reads through the silence.

I could see that Mr. Moran was very angry. The seventh-graders looked up, suddenly sensing danger.

Misha and I stared now. There was no looking down.

Mr. Moran touched his fingers one by one, and counted off. Boz was fighting for his case. Suddenly, Boz stopped.

We watched.

Boz began to cry. He tried to stop, and that made him shake a little bit. Mr. Moran said something else.

Boz came to the door, and walked out. We looked up at him.

He gathered his bookbag and turned away from his terminal. But just before leaving he looked back at Misha and me for a long time. No one said anything. The seventh-graders stared.

Boz left the room.

I lost my love of the room the day Boz cried. All of us did. We'd been given a choice, and as I looked back on what happened, I realized that Mr. Moran had treated us as adults rather than children. Boz was punished, expelled from the room for anything but necessary work relating to studying for the AP exam. No one in the administration was told of Boz's crime. There would be no blemish on his school record. After that Boz and I rarely spoke. He did not know who had told on him, and he would not ask. Nor would I tell. At times I comforted myself with the thought that my telling Mr. Moran, my emphasizing that Boz had meant no harm, perhaps had saved him. Still, I felt a certain ruthlessness and completion with what we had done. We were becoming adults, and our world had fair rules. Boz had violated them. That spring Mr. Moran made Misha and me Super Users. We had the right to a single digit. Where once I was 186,19, I became 186,0. Misha was 186,1. The single digit after the comma signified our new status. We could look at any password in the 186 group, including each other's. We were granted the technical term PROMAN, for Project Manager; the PDP, in a sense, had become our personal computer.

As adolescence began to recede, my obsession with computers started to shift. The absolute devotion and the joy of exploring new systems for their own sake were fading. That a system was new and unexplored was no longer enough; there had to be something more. There had to be a meaning that mattered in the real world. My move to adulthood coincided with the tectonic shift in the nature of computers: from a privileged hobby, an exotic terrain of Ataris, Apples, Commodores, and TRS–80s to a vast homogeneous sea of IBM-PCs and IBM clones, dotted with islands of eccentric holdouts. Arcades across the country had closed; a fad whose day was past. Games were categorized now. Just as horror movies, comedies, action/adventures, and romances were pitched to a particular audience, now games came in similar packaged flavors. A lexicon of video game design had been created, and designers played variations on the same hits. Atari went bankrupt and disappeared. A company called Nintendo had taken over, and they didn't care about programming. Theirs was a closed system. You couldn't program new games on their machines. You bought them off the shelf. Programming had become harder too. The IBM-PCs had 512K of RAM, and blazed at 16 megahertz. My Atari, with its 48K and 1.7 megahertz 6502 chip, was anemic.

As computers became more powerful, it was harder for kids to get beneath the glossy overlay of the images on the screen, down into the actual controls that made them work. We felt that loss. But worst of all was the feeling of the fun disappearing. Companies were promoting the idea of "home offices" and computers as tools for work. No one talked much about computers as tools for creativity. That original place beyond the screen, where we'd found so much extra life, appeared to shrink. The once exclusive shop had become a wide mall as millions of people discovered these machines. Our small culture had seeped into mass culture.

I'd been too busy to notice how growing standardization had trickled into the computer room, into the system room itself. Throughout eleventh grade we'd never needed to write any system patches. The big acquisition had been a set of color monitors, and these were disappointing compared to the vivid color I'd always had on my Atari. More subtle still was how each update of our system required less and less labor to install. The job of PROMAN, Super User, had been to administrate these tasks, which Mr. Moran, a full-time teacher, could not handle. It was a marvelous symbiosis between teacher and students born from necessity. So when I returned to the computer room in my senior year, I found myself with no class to take and nothing much to do. The only duty left was monitoring the seventh-graders, and this was not much fun at all. Even this role was waning. A new computer teacher had been hired to assist Mr. Moran. The demand for computer courses continued to grow. All seventh- and eighth-graders, along with new ninth-graders, were required to take at least one computer class. With the extra teacher, Mr. Moran could schedule things so that one of them was always there. But even with all these changes, the biggest change was within myself. I'd been elected Student Body President, the position Paul Hilal had once held. There was a wide world outside the room waiting for me, between my new duties as President, and Alysia. I'd finally found a girlfriend, and I wanted to spend my time with her.

*   *   *

In December of my senior year I was admitted to Harvard. With classes less important than ever, school began to resemble an off-peak summer camp. I'd come in and spend most of my day socializing, with classes serving as breaks in between. I stopped by the computer room less and less. Every once in a while I would see Misha inside, working on a program. He was there for professional reasons. For him the PDP became a place to do client-related work, editing programs on the school computer, which he then could deliver to his customers. Boz and I rarely saw each other. For him, the void left by the PDP had been filled by the analog world of set design. Racks of lights, carpentry, electrical wiring were his tools now.

A decade later I would find Boz living on the West Coast, where he worked as a full-time set builder for big-budget Hollywood films. We talked about the room.

"The fact that it was a multi-user system was really important," Boz said to me. "A lot of kids then didn't get any experience with ideas of privilege and protection and the whole idea of how a mainframe could work: how one microprocessor could manage all these projects." We agreed that our PDP had been a cyberspace of the best kind: a medium for collaboration, for shared experience, a social space. We'd had that unique cyberspace in the physical space of our computer room, a doubly social space, both online and off.

"Either you or Misha told Mr. Moran," Boz said toward the end of our conversation. I found myself mute, unable to confess, suddenly ashamed. "I was devastated," Boz added. For years he'd stopped programming, stopped using computers that way. "But lately," he went on, "I've been teaching myself to program in C++. It's strange, but I recently had this desire to do that." At night, Boz told me, he would sometimes stay up late and code for fun while his girlfriend slept.

I'd found Boz and made a kind of peace with him, but I never did find Misha. The last time I saw him was at our high school graduation. Standing in the sticky sun in my black robes that day, my mind was not on my friends or even myself but on my sister. Samantha had run away from home. My mom and dad were wrought with pain, terrified that she would be lost forever. They hired a private detective to find her. Within two days he found Samantha, living with Brian across the park, on the West Side. My parents lured her back with the promise of letting her do her own thing, a promise, it turned out, that was a ploy. Unbeknownst to me, my parents had made a plan to nab Samantha and deliver her to a special school in Maine that specialized in kids like her. Whatever they did, they did it right. Samantha would graduate from high school and go on to college, leaving Brian and the streets in her past.

I got into a huge fight with my mom just after graduation. At issue was whether I had to follow the rules in the house, which basically meant treating my stepfather with respect rather than with my customary bitter condescension. It was a silly fight, but one I instigated, on automatic pilot.

My stepfather interfered, and I decided to bring it all down, taunting him.

"Well if you don't like how I am, then I can leave. I can just move out and live with my dad!"

My mother stood between us, about to cry.

"Move out then," he said. "Move out."

"Fine! I will!"

And I walked back into my room and started packing, tossing clothes into a suitcase. The window across the courtyard was dark.

My mom followed me. She sat down on my bed, crying hard. "Don't do this, David," she sobbed. I just kept throwing stuff in my bag.

"I'll call you, mom," I said. I zipped up the bag.

In the closet was my Atari, with all its disks and cables and years of magazines, of code from *A.N.A.L.O.G.* and *ANTIC*, and copies of Cheese and my term paper and everything that I had ever built and dreamt on that machine. When I walked out the door I wasn't thinking about it. And I wouldn't think about that machine again for a long, long time. My mother watched me as I waited for the elevator. Later, much later, I thought about that moment. How both her children were gone and there was nothing she could do.

11

Hide the Geek

THE MAINFRAME IN THE SCIENCE CENTER at Harvard filled a room three times larger than the old computer room at Horace Mann. Through the glass I could see the elite users, the system managers, loading great spools of mag tape and tinkering with racks of modems. I stood in line with twenty other freshmen, idly watching as we waited our turn to take "the computer part" of our mandatory Quantitative Reasoning Analysis test, or QRA. Like Horace Mann's administrators, who years earlier decided to invest in a computer curriculum, by 1986 Harvard administrators grudgingly acknowledged that knowing something about computers was now necessary for well-rounded graduates. In the 1980s algorithms, in simplest form, had joined Shakespeare as part of the canon. The QRA was Harvard's way of demonstrating this: by testing our ability to build a conditional loop using an IF-THEN statement and a GOTO statement in BASIC. I secretly was looking forward to this silly test. I hadn't programmed in almost a year and I found myself missing it. Even a little IF-THEN statement would be fun to write.

Sitting at the terminal, the familiar monochrome pulsating, the comfortable feel of a keyboard under my fingers, I started to code, following the simple questions on the paper beside me. In a few minutes I'd built all the loops they wanted and passed the test. I'd come to Harvard thinking of majoring in computer science. As I walked out of the room I decided to investigate the possibilities. Paul Hilal, my idol at Horace Mann, was now a junior at Harvard. I decided to call him up. The great gulf of two years in high school was foreshortened in college, and Paul felt more like an equal than an untouchable.

Over a beer the next day, seated in a dark pub filled with wooden booths, the tables gouged through years of students having carved their initials, we talked about old times. "I have my own company," Paul told me. "We design software for studying the effects of drugs on the nervous system." Paul worked with his brother, and their clients were mostly pharmacologists and neuroscientists. They were selling thousands of dollars' worth of software, but it was difficult. Paul worked insane hours, between his business and his classes. It seemed natural though—that he would use his computer skills to build his own company. Paul was twenty, but already he had the wisdom of a seasoned businessman.

"You know," he said, "this is it. Everything is going to be computerized. The whole world is going to be using computers. We have something. We can get in now, and be part of this. I mean look at Bill Gates. He dropped out his sophomore year to start Microsoft. He found a niche. We can too."

"So what's going to happen with your company? Are you going to drop out of school to run it?"

"No. It's not the right kind of niche. It doesn't scale well."

I didn't understand what he meant.

"Medical equipment," he explained, "requires too much customization for each user. The more people buy my software, the more programming I have to do. What you want is a product that can work off the shelf, so as demand grows, all you have to do is ship, and ship, and ship more disks."

We batted around ideas for software that would scale, and this led to a discussion about the future.

"One day," I said, "we're not going to have personal computers anymore. We're just going to have these fiber-optic networks, and everything is going to come through there. And we'll be able to broadcast too. It'll be like having as many television stations as there are phone numbers. We'll all become TV producers."

Paul told me about a program at MIT called the Media Lab that had begun thinking about this. He said that Amy Bruckman, the only girl Super User at Horace Mann, had gone to MIT. Maybe, I thought, I could do something like that at Harvard. Study culture, history, and computers all together, and build software related to that, whatever "that" is.

"You could try the history and science department," he said. "Maybe you could do some kind of independent study through them." Drinking my beer, I thought this sounded like a great idea. Study the best things in the world to study—history and computers; why not?

"Computers are not a science," the department advisor said to me as I sat in her tiny office in the history and science department. "Computers are part of engineering, and engineering is not science. Science is physics and chemistry. And in some cases biology. Students in our department are expected to choose one of these three scientific areas."

"But can't we make an exception? I mean, computers are ultimately philosophical machines. They are based on logic. Isn't logic part of science? And math?" I suddenly realized that no one had ever asked this woman these questions.

She smiled, shifted around in her all-beige suit, and gave me a look that said "Freshman, you are not worthy."

It was like seventh grade all over again.

"Perhaps," she said, "You should go to MIT. There you can study computers the way we study science."

"But what about the history? MIT doesn't have the history departments Harvard has. The point is to integrate the two."

"Well, you can't do that here."

As I walked back to my dorm I thought about the second derivative and how all of this related to it. In calculus the derivative is a value that explains the rate of change of an object through time. For instance, a graph of a car accelerating from 25 miles an hour to 30 miles an hour has a derivative. The first derivative is the actual rate of acceleration—how fast its speed is changing—which helps us understand what the car's final speed will be. The second derivative, though, is much more interesting. It's the rate of acceleration of the rate of acceleration: the part that tells us whether the car's acceleration is about to change from, say, "fast" to "faster." If computers colliding with culture was an idea accelerating into public consciousness, then my meeting with the advisor had been about whether we both agreed on the second derivative—the hidden value that showed how fast computers were breaking out, leaping beyond the pond of engineering into the ocean of culture. There was no doubt in my mind that the second derivative was banking along an exponential growth curve, headed straight for infinity. In 1986 the symptoms of such change were far less perceptible than from today's vantage point. Computers had made a small leap, out of math into business, with a much-maligned detour into video games. These could easily be dismissed as "video nicotine" for the kiddies, craftily delivered to produce maximum addiction and idea-cancer in attention-deprived teenagers, whose emotional retardation is caused by lack of human contact. For that woman in her cozy office, with her Harvard letterhead and stacks of typewritten letters, I was one of those video-game kids, washed up on the beach of the academy, disoriented, grappling for the familiar.

Despite my disappointment I could see telltale signs of a change ahead. At Harvard and other elite colleges kids were lining up to buy computers. Two years earlier Apple had released a new machine—the Macintosh—and at Harvard we all received flyers from the company offering a 30-percent discount on their machines. My class, the class of 1990, was the first to attain the stratospheric statistic of 90 percent: Ninety percent of our freshman rooming groups, the school newspaper reported, had at least one computer among them. Much of this had to do with the Macintosh, which like some fecund genus of plastic mushroom seemed to sprout up everywhere—under piles of papers on messy desks, shoved beneath beds, between piles of dirty laundry, smeared with fingerprints, soda stains, and bumper stickers. The Mac did what no computer had done before—it made the artificial barrier between humanities and science disappear. It was all right for an English major to own a computer, if it was a Macintosh. That's because the Macintosh was designed for writing. And for the first time, kids who'd sneered at the geeks in the computer room, who'd looked down at computing as a vile nerd enterprise, lined up to get their 30-percent discounts.

For weeks my roommate waited to get his Mac. The machines were oversubscribed and the dealer had run out. New orders took a long time to arrive, so that by the time my roommate came back carrying a big box with the word *Macintosh* on it we'd all become curious—all five of us in our rooming group. I'd seen Macs before in Manhattan computer stores, but I'd never had the chance to do that quiet, private thing with a new computer that I liked to do, when no one was looking. It's like heavy petting. Or an autopsy. I like to fondle a new machine, strip away its layers, get down to the core and see what's inside. Any computer will have its top level of operating system and user interface: seeing how that connected to the inside of the machine by taking the time to investigate, to probe, can't be done in the showroom of a Midtown store with sales clerks hovering and customers behind you, waiting their turn at the machine. When my roommate brought back his Macintosh we crowded around as he set up the machine, which wasn't difficult: all you had to do was plug it in and connect the keyboard and "mouse."

Where's the operating system? was the first thing I thought as I stared at the Macintosh's screen. *Where is the system?*

There's no operating system on the Mac. Then I understood, with a mixture of bliss and disgust, that the interface *is* the operating system! You can't "go" deeper. This machine wasn't built for programming. It was built for using programs. It was a weird, disorienting feeling, seeing a machine that's so beautiful yet so remote, so secretive, that I couldn't get inside it. Later that evening I sat down with my roommate's Mac when no one was there and played for a while. The "icons" were beautiful. The "mouse" was intriguing. The fonts, onscreen, were extraordinary. I could actually see what my paper would look like when it was printed. But beneath these jewels lay a great gray void. Where the computer used to be, there was nothing. You could use this Macintosh—no, you could master this Macintosh—without having to understand how it worked. Magic, beauty, metaphor had replaced the tactile thing, the sinews of logic gates and charged electrons. I remembered the IMSAI 8080, with its front panel so deliciously close, and my Atari with its PEEK and POKE commands, thrusting me right into the stack, the very core of the machine. How far away and useless that knowledge seemed as I sat in front of the Mac.

On the surface the Macintosh appeared to represent the triumph of the ideals of a collaborative man–machine symbiosis that began in the 1960s, when computers created the first hacker cultures in universities. Here was a machine designed to "augment" our intellect through easy yet powerful software applications. Clumsy draftsmen now could draw perfect shapes with MacPaint. Poor typists could produce immaculate, professional-looking documents. Spreadsheets could transform even numberphobes into bean counters. The Mac widened access to computers by standardizing the operating system with a consistent visual look and feel, what people would come to call a graphical user interface. There was something vaguely psychedelic about the machine, and New Agey. Advertisements for the Macintosh were illustrated with high-quality computer graphics and "bit-mapped" fonts that looked on screen similar to what they looked like on paper. The ads implied that Macintosh was a better tool, a sharper, finer instrument than an MS-DOS machine. Macintosh was about self-expression rather than brute force calculation, like the IBM-PC; it also was about individualism and self-discovery. Here was the shaman with a "happy Mac" face on the startup screen: guide to understanding, enabler of creative dreams.

Although at that time I couldn't put my finger on why I was so disappointed with the Macintosh, I later came to realize what was missing was symptomatic of personal computers: gone was the feeling I'd had of a shared social space around a computer. I didn't know other people who felt that way, but then few of my peers had grown up with hands-on access to a time-shared computer like the PDP. Personal computers were inward-looking machines, meant to be used solo. Ironically, though I did not know it at the time, the very thing that made the Macintosh so exciting for new users—its graphical interface—had first been invented through research on creating a nationwide computer network, a continental version of our little PDP.

The same division of the Defense Department, the Advanced Research Projects Agency (ARPA), which had funded the creation of the Internet, also funded the first graphical user interface; they were meant to be one. The network would connect the people, and the people would communicate through a new system of "icons" and "windows" using a "mouse," navigating through a primordial information space—cyberspace. The prototype of the system, called NLS, for oNLine System, was unveiled in the fall of 1968 in San Francisco by its creator, Douglas Engelbart, after five years of ARPA-funded research and development to an audience of 2,000 in the city's Civic Auditorium. The audience watched in fascination as Engelbart clicked and moved through windows on the screen, his actions beamed to a wall-sized screen above, an astronaut from the future. Engelbart's interface was the gateway to the network, the visual metaphor that would turn an otherwise arcane breakthrough in wiring computers together into an information utility usable by all. Yet in 1986, nearly twenty years after the demonstration of NLS, such ideas seemed strange. Computer networks were kept separate from most people. Hobbyists and kids might use them to download software from bulletin boards, and corporations might use them for business, but the "ease of use" that would occur when the graphical interface met a nationwide network was still a dream.

The Macintosh marked the arrival of the Power User, the rise of the computer-literate autodidact who "knows" how computers work. It also changed the way people learned about computers. Where once learning meant teaching children how to program, the new breed of machine spawned a new curriculum that taught them how to use software. The Macintosh unleashed a torrent of product placements in the classroom, a fantastic boondoggle in which a new generation would grow up "learning" computers through a peculiar form of teaching: hands-on introduction to commercial software, during class time, at school. Lost was the idea of hands-on access to the way the machine worked and a curriculum rooted in collaborative programming. In our computer room, learning how to use software had always served as a gateway. First you play games, or produce hi-res images; later, if compelled, you can go below the surface, learning how these programs work. But in this new environment, encumbered by obstacles, students had less opportunity to take the next step.

At Harvard that freshman year I thought about how computers were supposed to catalyze a new culture, a new world. In a world built on information computers would devolve power, breaking hierarchies and hidebound institutions rather than further centralizing them, as had been imagined in the days when IBM mainframes ruled supreme. With cheap computers everywhere we'd all be citizen-hackers, each able to rip apart systems, understand their function, and build better ones if we chose. Systems built on lies would be harder to maintain and fooling people far more difficult. But it all felt dreary and compromised that fall. Resistance was futile. The world would never change. Computers were merely being assimilated, becoming another part of the system. In the computer room or at home alone with the Atari, copying pages from *ANTIC*, creating code of my own, my friends and I had been citizen-kings of the digital world. We were building something. Where other kids were watching TV and seeing movies—consuming media—we were creating media. We'd broken the system, gotten around to the other side and become producers in a land where, more than ever, kids were expected to be consumers.

All this was lost on most grown-ups in the late 1970s and early 1980s; strangely, they saw computers as corrosive to our literacy, mere mental junk food rotting our synapses, leaving gaping cavities in our brains. Mr. Moran knew better. And so did my parents. They'd bought me my computer and always supported my love of the machine. They saw the tangible effect the computer had on my ambitions at school and how it enabled me to become a better student. This was the computer revolution as I knew it, and it was supposed to spread out, get bigger, touch everyone in the world. But by the late 1980s that world seemed to recede, fade, replaced by yet another permutation of one-way media: digital consumption, computers as collectibles. Computers had become fashionable, and it came to be said, "You can never be too thin, too rich, or have too much RAM."

Thwarted in my intention to study computer science and history at Harvard, I decided to major in history and literature. As a sophomore in the history and literature department I got to study systems the old-fashioned way—through books and lectures. My fantasy from high school of World 2.0, with its spinning globe and embedded sets of data, came to seem nothing more than a vague metaphor, rarely contemplated. I didn't realize how much I'd buried and forgotten until one rainy afternoon as I wandered past a software store in Harvard Square a few blocks from my dormitory. Through the lines of water trickling

down the store window I saw a poster: BEYOND ZORK—
the letters crashing through a brick wall, surrounded by a
golden burst of light. Beyond Zork. Zork was the name of
the home computer version of Dungeon and the name of
the long-gone empire whose ruins I'd quested through on
the PDP. The misty rain clouded my glasses and dripped off
my nose. But I didn't care. For a moment, the mailbox and
the white house on the hill were there again; the heavy rug
and the lantern, the twisty maze of passages all alike, and
the magic password, *xyzzy*. Laden with books and
notepads, homework and classes, my backpack over one
shoulder, all I wanted to do was slip into the world of the
white house, to go beyond Zork. I went into the store and
bought the game.

My roommate Tom came home later that day. He was
from Reno, Nevada, and had a garrulous, rambling charm
that I imagined came from the great open spaces he'd
grown up in. He was an actor, an English major, and an extraordinary flirt. Girls compared him to Sam Shepard. As I
would soon discover, Tom and I shared a little secret.

I was at the far end of our living room, where we had two
desks back to back, seated at my computer. He strode in,
his cowboy boots tracking street dirt, and suddenly
stopped. Frozen. He stared down at the coffee table. "Hey. I
can't believe you got this," he said. He held the Beyond
Zork box aloft.

"I was so into Zork," Tom said.

"You were?" I was surprised.

I understood then that he too had once been a computer
geek. An Apple II kid. But he'd buried that uncool legacy so
deep that it didn't unfurl, in its full splendor, until that day.

Tom pulled a chair over to my desk. Six hours later we
were still playing Beyond Zork. We had trouble with the

organ grinder and his hurdy-gurdy. The sound was lethal. That was a tough puzzle, tougher still than the problem with the Mother Hungus.

"Mother Hungus?"

MOTHER HUNGUS! MOTHER HUNGUS! MOTHER HUNGUS! Tom and I suddenly were jumping up and down, screaming "MOTHER HUNGUS! MOTHER HUNGUS!" Six hours of computer gaming could do that to you: "MOTHER HUNGUS! MOTHER HUNGUS!"

For three weeks Tom and I played Beyond Zork together every day. People playing together through a computer. That's fun. That's different. In college I'd become so accustomed to the computer as a solitary thing, a private place like a book, good for one reader at a time. But used to play a game with someone, the computer became something else: a communications device; a strange social glue. What if we could get hundreds of people playing Beyond Zork together at the same time, each of them in the game? Tom and I both were reading *Neuromancer*, a sci-fi book by William Gibson, who wrote about a fantastic hallucinatory future and a world whose people were connected by a network he called cyberspace. In that book Gibson's network created beautiful visual reflections of the real world. The characters "jacked in" by linking to the network through their brains, experiencing an eerie feeling of leaving the body, flying, floating in another place built by binary data, digital representations of information that take on physical form. Beyond Zork is like that too, except words on a screen create the illusion of slipping through to another world. But the idea is the same; it's just low-res versus hi-fi. It was Gibson who made it clear, as I sat with Tom playing our game, that I'd been there. I'd been to cyberspace. That was our PDP. Me, Boz, and Misha. All those Super Users before us. We'd all come of age there, jacked-in through the text on the screen, synced up through our time-shared system.

Unable to find the right key for the scarecrow, Tom called the Beyond Zork Help line for us, pretending to be my dad. "I need a clue," he said into the phone. "My son is hyper because he can't get the key from the scarecrow"—he paused while the voice on the line said something—"I don't know what that means. But can you tell me how he can get the key from the scarecrow?" This was against the game maker's policy. You had to pay for clues by buying a special book with "InvisiClues" that were revealed by rubbing a special pen across the pages. Tom was such a good dad, though, that Technical Support caved and told him whatever we wanted to know. Muffling ourselves, biting our hands, the hardest part for Tom and me was not to laugh and give it all away.

Hide the Geek. That's what it was about. Adulthood was about hiding the geek. And reclaiming it. I'd been hiding it at Harvard. But in our room Tom and I could get it back. We finished Beyond Zork after finding a bug in the game that gave us unlimited gold coins so that we could buy all the best armor and win without solving all the clues. Then we got into SimCity after that and flight simulators. The point was to play together. That was the key thing. Together.

*   *   *

People using computers together as part of a network, rather than side by side at the same desk, was an image in the back of my mind two years later, in my senior year, when I had a job interview with Microsoft. When the big companies came to Harvard recruiting, I dutifully signed up for interviews with Morgan Stanley, McKinsey & Company, Goldman Sachs. Tick, tick, tick. I just checked off the boxes on the interview request sheet. Harvard assigned us each 1,000 "points" to bid on interviews. Going through the stack of companies, I saw one I recognized: Microsoft. I put 40 points on them. A few weeks later I received a letter with a list of companies that would interview me. Microsoft was on the list. So was Morgan Stanley and McKinsey. Microsoft was the only company not in banking or consulting.

"Harrowing," I told my dad on the pay phone on the second floor of the Harvard library, when he asked me how it was going. Next door, fifty seniors and recent graduates were playing musical chairs in a big room. They faced representatives from companies across America. It was depressing, I told him. I didn't know anything about Wall Street or "consulting" or banking or this thing called "marketing." I wasn't sure what I knew. A lot about history. How to cheat in Zork. Linked lists, packed arrays, RSTS/E, player-missile graphics. I had to hang up. My last interview was with Microsoft and I saw that it was time to go. Then I could go home and take off my suit.

*   *   *

"Why do you want to work at Microsoft?"

"It's all about fiber optics," I said, cryptically. On the other side of a thin plastic table with simulated wood grain, the man from Microsoft slouched in his metal chair. He wore a plain blue T-shirt. All around us I could hear the murmur of interviews at other tables. Everyone else, interviewee and interviewer, wore suits. It was five o'clock, and this was the last interview of the last day for both of us. Tomorrow he was going back to Seattle.

"Yes?" he said, unslouching from the chair. He seemed young, maybe mid-twenties, with tousled blond hair and the healthy shine that comes from not living in New England.

"Yeah. Think about it. The phone companies are laying all this digital fiber. It can carry audio, video, information—it's all data. Once that's in place, the whole way we use computers is going to change. Software will change. It'll be about communicating, working in groups over long distances. New kinds of entertainment are going to emerge. It's going to be huge," my voice rose with excitement. "The hugest thing ever. What's Microsoft going to do about it?"

Microsoft. A joke really, I thought. This company makes MS-DOS and a word processor. They're not a game company. They don't have that intangible "it" quality of a hot software company. Bland is more like it. I felt reckless, and spieling about the future was a lot more fun than pretending to be interested in investment banking. The interviewer leaned toward me.

"That's true," he agreed.

I stared back. Huh? What did he say?

"Computers are not going to stay the same forever," he went on. He asked why I thought being on networks would excite people, and all of a sudden I mentioned Zork. Zork! In an interview? It just came out, and I couldn't take it back.

"I loved that game," he said. "It was the game that made me want to understand computers." We started geeking out, together. He told me about a networked version called Adventure! I couldn't believe it. Why hadn't I heard about this before? Tom and I could have played against each other! The man from Microsoft knew about *Neuromancer* too. *He gets it!* I thought. We riffed on the future, imagining self-produced television shows, home-brew media.

"How would you like to come to Seattle?" he said, all of a sudden.

"Yes," I said. Yes.

12

Beyond Zork

THREE WEEKS LATER I was flying to Seattle. Here, looking
down at grids made by country roads, highways, and rivers,
it was easy to imagine the evanescent world of cyberspace.
A power grid of information. I gave little thought to what
Microsoft actually sold or what their core business in 1990
was—selling MS-DOS, Windows, Word, and Excel. Instead I
fantasized about working in a tiny laboratory with artists
and programmers, building experimental environments,
sculpting the shape of tomorrow. As the plane descended
toward the Pacific, the afterburn of the sunset spreading
over Seattle, I stared out the window thinking how incredible it was that I'd found a place that understood the future.
A company that knew that soon all our computers would
be connected to one national, eventually global, universal
network, and that in the process everything we assumed
about personal computers would evaporate and be replaced
by a new environment, a network that would provide the
electronic earth for humanity to build the next iteration of
civilization. The joy would come from working toward what
we as a society would create. And surely, I felt, it would
lead to a better world, one where the tyranny of media
monoliths and advertisers tweaking consumer impulses
would no longer dominate the production and distribution

of television, films, radio, and print, nor shape our perception of events. In a land where reality comes from television, high-speed computer networks would usher in a smarter, juster age. Because we would control the new media. Most important, in the process all of us would contribute to the construction of the latest stage of human culture, with all its attendant richness and complexity. Some would simply vote with their fingers, switching over to homebrew TV; others would produce the art filling the fiber. I assumed Microsoft was hatching an in-house project to probe these questions, and I paid no thought to how this might be profitable. The idea mesmerized me, and I anticipated what would happen, fantasizing about secret groups of programmers and artists working in consort to explore the twenty-first century.

"Sorry I'm late," Debbi said, shaking my hand. "Come on upstairs." Debbi was my handler. She would coordinate all my interviews. "I hope your flight went okay. Did you have any trouble getting here? You know the numbers on the buildings can get confusing for some people." I assured her that I wasn't one of those people. The Microsoft campus map was perfectly legible, and the numbered buildings were no problem, I said coolly. As we walked up a series of carpeted beige stairs, circling the open lobby with a reception desk at the center, I looked outside the big front glass wall and saw acres of green lawns and low buildings—five or so floors—draped entirely in glass and dented like origami to give each building maximum window surface area. It was logical. Smarts uber alles, a controlled environment built by and for computer programmers. A hacker homeland! Standing on the landing of the next floor, Debbi mentioned the free soda. Microsoft employees didn't have to pay for soda. "Would you like one?" she asked. I said yes.

I knew something was wrong when Debbi, sitting in her little office (no window, lots of filing cabinets, presumably stuffed with résumés like mine), told me my first meeting was with Ellen in "marketing." Debbi grinned—as I did, to conceal my apprehension. Her frizzy black hair radiated a powerful scent of hair gel. *It smells like Prell shampoo*, I thought. "This is your schedule," she said, handing me a piece of paper. "My number is here. If you have any problems or questions you can call me. I know you won't get lost, but if you do, call me and I'll find you. It happens sometimes." The first interview, with Ellen, was in this building. On the same floor.

"Okay," I said. "Great. Thanks."

"Great!"

Shooting down the beige hall, following the numbers on the doors, it took a minute to reach Ellen's office. Ellen, clicking away at a keyboard and surrounded by product boxes for Microsoft Excel, stood to greet me. She loomed large, close to six feet, with an enormous black velour headband and huge round glasses. I felt the first of several tremors of anxiety. Ellen offered me a soda, which I accepted, and we moved to a conference room a few doors down. Minutes into our conversation I soon realized that not knowing what the word *marketing* meant was a major conversation stopper. I'd never spent an instant in an economics class, rarely read the business pages of anything, and had had summer jobs where *market* as a verb never made its way into conversation. As she explained what this marketing involved, I recognized that a terrible mistake had been made. Where was that guy in the T-shirt? The one at Harvard. How could he do this to me? Ellen was interviewing me for a job as a foot soldier to sell, of all things, Microsoft Excel. Feeling my pupils contracting, I segued into my speech about the future of media. As I finished Ellen attacked the telephone, initiating a flurry of jabbing phone calls to set up a meeting with the "right person."

The right person was not available until the next day. Debbi arranged to extend my stay at the Holiday Inn, which I thought boded well. The right person turned out to be a chain-smoking Dutchman. There was a very un-West Coast air in his windowed office. It was covered with white boards, scrawled with C++ code and diagrams. He was one of the elite programmers who'd created the latest version of Excel, which he queried me on, after offering me a soda. What do I think about the features as they are now, what would I change or add? Slightly befuddled, my drink untouched, unwilling to confess that the only Microsoft products I'd ever used were MS-DOS and their flight simulator game, I valiantly tried to play along until it finally became clear that I was not there because I wanted to program future versions of Excel. "What do you want to do?" he asked. I told him. The word *multimedia* triggered a connection. A few phone calls later (less jabbing than Ellen, more languid), and Debbi, through his phone, announced that I should stay another day. Where I really ought to go, she said, is the experimental new media group. That, I thought, sounded right.

Day three began like the previous two—a fine breakfast of extravagantly good, cheap, and plentiful pancakes, followed by loads of coffee and a quick drive to the euphemistically named "campus." In forty-eight hours there I'd learned that many Microsoft employees were shipped directly from college to the "campus" where, like hermetically sealed novices, they immediately set to work. But unlike the Harvard campus, this was a closed environment. As I'd started to sense, my conception of Microsoft as a company that made workaday software, the oatmeal of personal computing, had been initially correct. Why would they waste resources on an open, system-challenging research lab? Here the spirit of hacking was tamed, away from exploring systems to fine-tuning existing ones.

Debbi, I thought, had been assigned to me because she was Jewish and from Long Island. She took me out to dinner the night before and confessed how hard it sometimes was to be so far away from the East Coast. The Microsoft man at Harvard seemed more and more to have sprung from nowhere, a gremlin of my imagination. Maybe he'd had no understanding of what I'd been saying that afternoon and simply wanted to fill his quota of recruits to send West. I was tiring of the campus with its straggly disconnected architecture and rootless employees scurrying by in open-toe Birkenstock sandals from office to free-soda machine. "Look," Debbi had said to me the day before, on our way to dinner, pointing with pride, "there are people having a meeting outside." Across the way, on a mound of grass between two polygonal buildings, were a cluster of people drawing diagrams with a portable whiteboard. "It's like having class outside," I said. "Exactly," Debbie replied, pleased that I saw the connection to the ethos of free-flowing, group exploration.

Mid-morning on day three I was brought to one of the older Microsoft buildings. Plainly rectangular, a relic from the pre-windows maximizing phase. There I was greeted by the first hacker I'd seen at Microsoft. He had a big, bushy mane of curly black hair, black jeans and T-shirt, basketball sneakers, and pitted skin. He said hi without looking at me and took me to his office, which looked and smelled like a cave. The lights were off, and several computers glowed in the gloom. Before allowing me to look, he asked if I knew about the free soda.

"Yes," I said.

"You want one?"

How much more Coke could I drink? It made me pee. The entire trip had been punctuated by bladder anxieties. Holding tight, waiting to get to the bathroom between interviews. But I couldn't insult my hosts by refusing their hospitality.

"Sure."

He escorted me to a familiar sight: a windowless room, fluorescent lights aglow, ringed with vending machines that distributed sugars and caffeine. After we popped open our Cokes he asked, "How would you redesign that machine?" pointing to the Coke machine. It was essentially a big fridge with a Coke ad, a lock, and a money receptacle. Perfect. Why would anyone want to change it? But he expected an answer, so I thought a minute and then told him I'd connect the machine to a telephone line so the vendor could tell when the stock was running low. Coke would never run out that way. Maximum sales.

"No," he said, "not the back end. The front end." I nodded. "How would you redesign it?" Suddenly I had visions of nightmare college interviews, where they ask, "If you could be a vegetable, what kind would you be, and why?" The Coke machine here probably delivered more Coke per day per person than any other Coke machine in the history of soda. Sales were maxed out. So who cared about the "front end"?

"I would put a flat-screen LCD panel on the front here," I said, pointing to the front of the machine, "and run ads for the sodas inside. I would run promotions based on volume. Say the Sprite wasn't selling; I would discount it. I would put a keyboard in, or a touch screen, and ask questions, have questionnaires built into the machine and if you took the time to answer them you could get a free soda or maybe a discount." That seemed right. Geek out! But the further I went with this the more impatient and disgusted he seemed. There was something I wasn't getting. Some Valhalla of geek logic that eluded me.

"Let's go inside," he said, nodding toward the door. Inside, I guessed, was where his office was. This room must be outside.

Back in his office, staring at three monitors side by side, I watched as he showed me the latest developments in the lab. This was the end of the line, the last possible place for me at Microsoft, and I was eager to see what projects he was working on.

"We put multimedia into the entire range of Microsoft products. Multimedia enhances the user's experience, and makes our products easier and more enjoyable to use." He took the mouse and clicked open a window in Microsoft Excel.

"See here," he said, indicating the monitor. I peered at it. The electronic buttons at the top of the screen seemed different. Rounded. Shadowed. Not flat and two-dimensional, as they usually were.

"Nice," I said.

He clicked on a button with his mouse; from two stereo speakers emanated a very satisfying metallic *click*.

"Nice."

"This is something we're working on now." He clicked on the scroll bar at the side of the spreadsheet window and slid it downward. As we moved down the spreadsheet, his stereo speakers made a deepening sound, like a note sliding down in octaves.

"We're going down," he said. "We're going up." As he slid up, the speakers ascended in octaves, getting brighter.

"Ahh," I said.

"Tactile feedback," he said.

Six hours until my plane took off. Maybe I'd have enough time to drive around Seattle in the afternoon. Suddenly I noticed the silence in the room. The speakers weren't whooshing anymore.

"That's great," I said, realizing he expected me to say something. I had to pee again. Maybe this would be a good time to ask to use the bathroom.

* * *

I never got to work with Microsoft. A few weeks after my return to New York I received a letter explaining that there was no suitable job for me. I'd been at a place dedicated to fine-tuning the surface of software. With their perpetual upgrades, each existing as a stratum of opaque sediment, further hiding the machine, burying it in ossified layers, there was little incentive to break out in a new direction. They were a short-term shop in a computer world where short—three months—seemed long to those mired in the latest code update. This wasn't evolution—it was fossilization. The trip to Seattle had paid off, however, in a way I didn't expect. It reawakened my excitement about computers. I realized that in four years of college I'd lost contact with something I knew and loved: programming. I took a freelance job writing databases, and when my stepfather retired and closed his advertising agency he gave me one of his old Macintoshes. At home, in my one-room apartment, paid for by my revivified skills at designing arrays and linked lists from the time of Cheese, I connected a modem to my computer and discovered an eerie parallel world to the PDP and RSTS/E called the Internet and UNIX. The first week I logged on to the Internet, using text-only commands and a program called Telnet, I felt I had arrived. Had come home to my familiar. The white mailbox at the foot of the hill was gone. But now there were thousands—maybe millions—of people gathered, building, doing, creating something out there, wherever *there* was. A once secret, hidden universe, a kid's world—a geek's world—wasn't secret anymore. And as more people got into it, I could feel the system reaching out to the point where it would cross over and become a rich part of our culture, world culture, equal to and perhaps one day beyond television, film, books, magazines. A new media. The second derivative had become the first. Here is where it would happen—not at Microsoft, or in some mysterious research lab. It would happen out here, built by us, for ourselves.

*   *   *

In the spring of 1996 I found a letter in my mailbox. The paper was thick bond, and my name was written by hand in big blocky letters. It looked like a kid's writing. I rarely get handwritten letters on paper; most of my mail is electronic. Pushing aside the usual incoming detritus of bills and catalogs, I sat down at my kitchen table and opened up the strange envelope. "Dear Mr. Bennahum," it began, "I am a seventh grader at Horace Mann and I would like to invite you to come to school and be interviewed for my seventh grade history project. We are interviewing alumni from the school who are doing things we are interested in. I am interested in computers and the Internet and would like to interview you about that." Enclosed was a second note, from his teacher, inviting me to call her if I had any questions. I phoned the teacher.

"Henry is a very shy boy," she told me. "He likes computers and he's very smart. It would be wonderful if you could come to school." School? The idea was delicious. Wake up at 6:30, get on the Number 1 train to the Bronx, walk up the hill. Why not? Maybe the pizza place would still be there. What video games would they have now? The PDP. Mr. Moran.

"Okay," I said. "I'll come up."

On the train I don't do math homework. I read the paper. But I ride in the first car, and the kids are still there. Doing their math homework. Playing around. To them I'm invisible, a grown-up. Just a guy reading the news. At 242nd Street I get out with the kids and head down the metal stairs to Broadway. The pizza place is still there. Inside I find games, and already some kids are there playing. Mortal Kombat, Pole Position. The walk up the hill takes five minutes. At the top it looks the same. The cafeteria, the main building, the football field. Slipping through columns of kids with schoolbags, I come to the seventh grade office to meet Henry. He's not there. I'm told to find his teacher. Maybe he's in class. His teacher is young, perhaps my age. To a kid, though, she'd seem old. I look through the window in the classroom door and she sees me. She tells the twenty students in her room to sit still for a moment, and as she comes outside to meet me I can see the kids leaning forward to look out the door, to see what is going on because classes are rarely interrupted.

"Henry is sick today," she explains, turning red. "I am so sorry. We didn't know until now. I am really sorry that you came all this way. Please accept our apologies." She is truly sorry. Like I'm out of sorts or something. Hardly. I am delighted to be here.

"It's okay, really it's okay. I've wanted to come up for a while anyway."

We agree that Henry will call me. We'll do it by phone.

"He'll be so disappointed," she tells me. "But you know, he is absentminded and I know how much this meant to him. So maybe this will teach him an important lesson about being responsible for things."

She seems like a very good teacher. I am sure I would have had a crush on her in seventh grade, if she'd been my teacher.

Crossing the green toward the main building, I look up to the third floor. To the windows. That's where I want to go. To the computer room. Classes have started, so the halls are empty. On the third landing I make a left and follow the shiny wood floors to the door, with its chicken-wire crossed glass. Inside I see fifteen kids working on Macintosh computers. I twist my head. The system room is still there. But there's no PDP inside. No Mr. Moran either. Just offices. Teachers' offices. It's the computer department's office. No command-line interfaces in there. No time-shared operating system. No software to be handwritten by the kids from scratch for the rest of the room. Not all boys either. Girls are in the room too. I go inside. It's a lab period. The room is hooked up to the Internet. Mr. Kenner, the new computer room teacher, invites me to sit in his office, in the spot where magnetic tapes used to be stored. As I come in I realize this was the spot Boz had stood the day Mr. Moran told him he'd never be Super User.

We talk about the changes. How the PDP was eventually carted out and thrown away, an enormous piece of junk, too obsolete to be sold. I picture it rusting under a pile of trash somewhere in a vast garbage dump. Do they have Super Users anymore? No, Mr. Kenner explains, the system is too decentralized. There's no need for a Super User. What about Pascal; do they still learn Pascal? Only some kids, the ones who want to study AP computer science. I start telling

Mr. Kenner stories, like the day Paul Haahr decompiled RSTS/E and figured out how the source code worked and then how he set out to write an operating system of his own.

"The kids who are like that now, they run our Web server. And they do UNIX."

"Don't you think something's changed?" I say. "If kids want to learn how computers work now, it's much more difficult. The system is too complex. You can't take it apart. You can't see it from the inside."

Mr. Kenner looks at me and smiles. "Some things can't be remembered forever," he says. "They're learning a lot of the same things though. Those who want to learn, can. They can still go deep. It's all part of a continuum."

On my way out Mr. Kenner points to a file cabinet. Maybe I would like these? I open the drawer and pull out a pile of yellowed green-and-white computer paper. The big sprocketed 24-inch kind. I feel dizzy for a moment. It's Spy programs. Spy programs! I see Amy Bruckman's name. Joel Westheimer's.

"I remember this program," I say, flabbergasted. "Mr. Moran held it up and showed it to us, saying it was the best one ever written." There is the grade Mr. Moran wrote in big letters across the top of Joel's program: "WOW!!! A+++."

"You can have them," Mr. Kenner says. "I knew we were saving them for all these years for a reason." He digs deeper in the file and comes out with a stack of blue notebooks. I get very still. The air around me feels sharp. AP computer exam notebooks? Mr. Kenner puts them on the table. They're from 1985. My year. I see Boz's book. Misha's. Mine.

"Can I have that?"

Mr. Kenner hands me my book. Inside, with some trepidation, I see that the handwriting describes Pascal programs. I trace the outline of the algorithm instinctively, the world of pure forms suddenly real. As I look up at the kids in the room, watching them—some are building Web sites, others are writing papers—it all seems like so much collage. A world of cutting and pasting. A place where the amphetamine high of pure form has given way to the treacly rush of toyland. Where once a hard game of Tempest gave me deep satisfaction, I could see that surfing a well-done Web page or animation gave these kids the same kind of pleasure. But where to go next? Games were my bait, leading to a wonderful switch into a realm where ideals were not ideals but the actual, tangible motor behind the code I held in my hand in that little blue book layered with pencil scrawls. Where else did such a world exist? The first generation to grow up on the Internet faces an all too similar danger. What first promises to be an extraordinary intellectual expansion is quickly, by virtue of its popularity, undermined by the market's inclination to reduce the sophisticated to the homogeneous.

It took the home computer a good seven years to be reduced to an opaque "black box," gussied up with layers of "idiot proof" interfaces. The Net, born a clear glass box, offers the intrepid explorer a rational system, clearly visible, open to the probings of all comers. The Net, like our junkyard-bound PDP, offers a system run by the people. Programs written and distributed for free. It began as a system of communal programming where one could build on the previous author's work, a place where the purity of form was the prime measure of an idea's success. This virtuous circle marks the original spirit of the Net and remains the soul of the network. But how long will this last? Already the forces of commercialization are threatening to choke off this primal source of inspiration.

The establishment of standards for future network services—such as multimedia—are fraught with discontent. There is a strong desire to own the patents and form the companies, which will in turn control the protocols and standards undergirding the Internet, thus ending nearly thirty years of shared group programming. The very quality that made the Internet alluring—its ability to connect myriad computer systems, low cost, and barrier-free structure— is at stake. Why should things be open or clear since private ownership, a killer-application of one's own, is the apotheosis of our age? The Net, with its brief, brilliant promise of a new media freed from artificial monopolies—a place where constraint and scarcity is replaced by abundance and liberty—faces the prospect of mutating into just another form of television, where ideas, distribution, and manufacture are all owned, trademarked, patented, and squeezed into neat corrals of property, all the better to derive a fortune. Looking at the kids in that room building their sites, I could only wonder if their words, programs, and pages were one more step toward a new global culture, a place where people build media and connections and take responsibility for informing themselves, rather than passively awaiting the consensus of the nightly news—or else a last gesture, from a time already past.

I knew that one of these kids, somewhere in the room, probing the sinews of code that make up the tissue of the Internet, would stumble upon an insight, a visceral apprehension of some deep truth, like the time I found infinity within a simple recursive loop. The benefit of our probings and self-teachings remains constant: to understand our symbiosis with machines. Our closeness to technology—

from the mundane answering machine to the hundreds of microprocessors in our cars—will only increase, and the temptation to reduce its underlying complexity to glossy simplicity will increase along with it. Such camouflaged technology changes into wizardry or sorcery, becoming the alchemy of our time, and all the brilliant insight into the rational forms underneath will mean nothing, forgotten to all but the initiated, those who aren't dazzled by the surface, and the few magicians with the power to define what is and isn't possible. To know the machine as we did, so intimately, is to forever change the way we experience our machine-mediated world.

It began as a game, one smooth quarter after another. It ended as knowledge. Somewhere in that great ganglion of logic we found a truth so bright and passionate that all our adolescent selves trembled to it, and we were changed. Where others found meaning in the poetry of Rimbaud or the lyrics of Jim Morrison, we found it in the layers of code and the perfect matrix of machine memory. Walking back down the hill, my old exam in hand, I thought of my old Atari. It now sits in my room, a memento mori, a tangible link to a fleeting point in time when the computer's own childhood, with its 48K of gloriously accessible RAM, matched mine, a time when with wild exhilaration we discovered a place all our own.

Epilogue

ERIC AND TIM ARE LIVING IN NEW YORK. After his stay in prison, Eric went to college and now works as a real-estate broker in Manhattan. Tim lives at home with his mother and works at a local espresso bar.

Scott Matthews runs his own Web site development firm, Turnstyle Web Site Architecture (*www.turnstyle.com/*), and was kind enough to give me his back issues of *Softside*, *Joystik!* and *ANTIC* magazines.

Aaron lives in New York and works as a film editor. He no longer owns a computer.

Amy Bruckman received her Ph.D. from the Media Lab at MIT, where she wrote her dissertation on MUDs—networked environments directly descended from Adventure and Zork and the days when you could FORCE hop, hop someone out of a game. She is an assistant professor at the Georgia Institute of Technology.

Joel Westheimer is a professor of education at New York University. At Princeton he studied programming in C for a while, but lost interest without the special environment of Horace Mann's computer room.

Paul Hilal enjoyed a meteoric career at Broadview Associates, a leading mergers and acquisitions firm specializing in high-tech businesses. He currently manages a technology sector investment fund.

Paul Haahr is the chief scientist and cofounder of Jive Technology in San Francisco, creating interpreters and compilers for the Java programming language.

Ian Allen works for a software firm in Cambridge, Massachusetts.

Misha Schwartz has disappeared. I was unable to find out where he is today.

Jeremy Bozza went to Brown, and now lives in Portland. He works as a set designer for Hollywood movies.

Mr. Moran left Horace Mann in 1988. He now teaches Windows NT, VMS, and UNIX administration through the Global Knowledge Network, an international program in continuing education. In 1985 Mr. Moran married Horace Mann's Latin teacher. A few years ago he told me that BIG-BRO and LILBRO did nothing, false sentinels meant to dissuade us from naughty deeds.

Samantha graduated from Hampshire College and went on to study fashion design in Paris. She's moved back to New York and has two e-mail accounts.

My mom and stepdad maintain their Web page on their own. My dad feels that with a little more work, he could become "really good at computers."

I live in New York City and write about technology and science. I have an Internet domain of my own, which you can visit at *www.memex.org/*.

Acknowledgments

I WOULD LIKE TO THANK Ted Byfield, Starla Cohen, Malcolm Gladwell, Larissa MacFarquhar, Douglas Rushkoff, Leslie Stevens—and Sarah Saffian, my doppelgänger—for reading early drafts of this book and providing me with essential feedback. Thanks also to Omar Wasow, for a particularly cogent discussion of how 8-bit programs programmed us as much as we programmed them; and to Miranda Dempster, whose curiosity and love were inspirational.

To the gang from the computer room—Jeremy Bozza, Amy Bruckman, John Fedlschuh, Paul Haahr, Paul Hilal, Scott Rosenberg, Joel Westheimer—I am grateful for your generous retelling of memories. And to Scott Matthews, who spent an afternoon with me digging through his collection of 8-bit memorabilia; Adam Kenner, for welcoming me and letting me feel at home; and Ed Moran, a great and generous teacher without whom this book wouldn't exist—thank you for sending me your textbook on RSTS/E (printed out by Super Users) and untangling the genesis of the computer room. Your password was OCHPAX (Joel told me).

I also want to thank the intrepid and dedicated Tina Bennett, whose commitment to this story was unwavering. You are simply the best agent on the planet. This book began with Judith Shulevitz, literary *shadchanis* extraordinaire, who first introduced me to Susan Rabiner at Basic. Thank you Judith. And Susan, thank you for showing me where the real story lay; and Eamon Dolan, my next editor, for inspiring me to go further; and finally, Jo Ann Miller, for gracing me with a firsthand experience of a master editor. You are a gift to publishing—pass on the skills to the next generation! And to Libby Garland, Marian Brown, and Matty Goldberg at Basic Books, for all your smart thoughts and support. Special thanks to Richard Fumosa and Michael Wilde for careful copyediting and shepherding this book to completion.

Martin Cohen, master teacher and advisor: your greatness and commitment showed me the way.

My stepdad, Edwin F. Lefkowith, for your love and enthusiasm. My mom, Christie Mayer Lefkowith, for everything.

My dad, Michael Bennahum, for reminding me how I got my first modem—and for being a marvelous father and my best friend.

Samantha, thank you for helping me remember; you are a great woman.

In Memoriam:
Muriel Bennahum, 1914–1991.